Shape Memory Polymer Composites

Shape Memory Polymer Composites discusses the fabrication of smart polymer composites and their material characterization. It covers shape memory polymer composites with two different types of reinforcement: shape memory polymer nanocomposites and shape memory hybrid composites.

Enhancing the mechanical and thermomechanical properties of shape memory polymers makes them an important class of materials for new-age applications ranging from aerospace, biomedical, electronics, to marine engineering. This book discusses how shape memory polymer composites exhibit remarkable mechanical properties, as compared to their corresponding shape memory polymers, without compromising their shape memory behavior. It presents experimental case studies of polymers, polymer composites, and multiphase composites, explaining the effects of each reinforcement on the material properties with corresponding simulation.

This book will be a useful reference for industry professionals and researchers involved in the mechanics of shape memory materials.

Emerging Materials and Technologies

Series Editor: Boris I. Kharissov

The *Emerging Materials and Technologies series* is devoted to highlighting publications centered on emerging advanced materials and novel technologies. Attention is paid to those newly discovered or applied materials with potential to solve pressing societal problems and improve quality of life, corresponding to environmental protection, medicine, communications, energy, transportation, advanced manufacturing, and related areas.

The series takes into account that, under present strong demands for energy, material, and cost savings, as well as heavy contamination problems and world-wide pandemic conditions, the area of emerging materials and related scalable technologies is a highly interdisciplinary field, with the need for researchers, professionals, and academics across the spectrum of engineering and technological disciplines. The main objective of this book series is to attract more attention to these materials and technologies and invite conversation among the international R&D community.

Polymer Processing
Design, Printing and Applications of Multi-Dimensional Techniques
Abhijit Bandyopadhyay and Rahul Chatterjee

Nanomaterials for Energy Applications
Edited by L. Syam Sundar, Shaik Feroz, and Faramarz Djavanroodi

Wastewater Treatment with the Fenton Process
Principles and Applications
Dominika Bury, Piotr Marcinowski, Jan Bogacki, Michal Jakubczak, and Agnieszka Jastrzebska

Mechanical Behavior of Advanced Materials: Modeling and Simulation
Edited by Jia Li and Qihong Fang

Shape Memory Polymer Composites
Characterization and Modeling
Nilesh Tiwari and Kanif M. Markad

Impedance Spectroscopy and its Application in Biological Detection
Edited by Geeta Bhatt, Manoj Bhatt and Shantanu Bhattacharya

For more information about this series, please visit: www.routledge.com/Emerging-Materials-and-Technologies/book-series/CRCEMT

Shape Memory Polymer Composites
Characterization and Modeling

Nilesh Tiwari and Kanif M. Markad

CRC Press
Taylor & Francis Group
Boca Raton London New York

CRC Press is an imprint of the
Taylor & Francis Group, an **informa** business

Designed cover image: © Shutterstock

MATLAB® is a trademark of The MathWorks, Inc. and is used with permission. The MathWorks does not warrant the accuracy of the text or exercises in this book. This book's use or discussion of MATLAB® software or related products does not constitute endorsement or sponsorship by The MathWorks of a particular pedagogical approach or particular use of the MATLAB® software.

First edition published 2024
by CRC Press
2385 NW Executive Center Drive, Suite 320, Boca Raton FL 33431

and by CRC Press
4 Park Square, Milton Park, Abingdon, Oxon, OX14 4RN

CRC Press is an imprint of Taylor & Francis Group, LLC

© 2024 Nilesh Tiwari and Kanif M. Markad

Library of Congress Cataloging-in-Publication Data
Names: Tiwari, Nilesh, author. | Markad, Kanif M., author.
Title: Shape memory polymer composites : characterization and modeling /
Nilesh Tiwari and Kanif M. Markad.
Description: First edition. | Boca Raton, FL : CRC Press, 2024. |
Includes bibliographical references and index.
Identifiers: LCCN 2023024118 (print) | LCCN 2023024119 (ebook) |
ISBN 9781032489940 (hardback) | ISBN 9781032490007 (paperback) |
ISBN 9781003391760 (ebook)
Subjects: LCSH: Shape memory polymers. | Polymeric composites.
Classification: LCC TA455.P585 .T59 2024 (print) | LCC TA455.P585 (ebook)
| DDC 620.1/92–dc23/eng/20230629
LC record available at https://lccn.loc.gov/2023024118
LC ebook record available at https://lccn.loc.gov/2023024119

ISBN: 978-1-032-48994-0 (hbk)
ISBN: 978-1-032-49000-7 (pbk)
ISBN: 978-1-003-39176-0 (ebk)

DOI: 10.1201/9781003391760

Typeset in Times
by codeMantra

Dedicated to our Parents

Contents

About the Authors

Dr. Nilesh Tiwari is a Management Executive in the Department of R&D and Quality Control at Sunrise Industries India Limited. He holds a PhD in Mechanical Engineering from the esteemed Sardar Vallabhbhai National Institute of Technology, Surat (SVNIT, Surat). His research is centered on the development and thermomechanical analysis of Smart Polymer Composites. He has a notable publication record with over 15 research papers in reputable journals, along with 8 articles presented at national and international conferences. He has also authored two book chapters and holds a patent. In addition to his industry involvement, Dr. Tiwari possesses over 5 years of teaching and research experience across various engineering institutes. His areas of expertise include polymer composites, multiphase computational mechanics, shape-memory material fabrication, and material characterization.

Dr. Kanif M. Markad is an Associate Professor in the Department of Mechanical Engineering at Dr. Vithalrao Vikhe Patil College of Engineering, Ahmednagar. He completed his PhD at Sardar Vallabhbhai National Institute of Technology, Surat, with a specialization in "Finite Element Analysis of Shape Memory Polymer Composite Beam." He has over 15 publications in reputed journals and has presented 7 articles at various national and international conferences. He has extensive experience as an academician and researcher in reputed engineering and technology institutes. He taught Machine Design, Engineering Mechanics and Theory of Mechanics for more than 13 years. His areas of interest include computational mechanics, finite element analysis, smart composites, and material characterization.

1 Introduction to Shape Memory Polymer

Shape memory polymer (SMP) has dynamic "memory" characteristics for shapes. It can transition from a stiff to a flexible state before returning to a rigid state. If left unrestrained, SMP will return to its "memory" shape, or it can be stretched, folded, or conformed to other shapes while being malleable. SMPs are compound plastic polymers that have a special chemical structure. Glass transition temperature (T_g) plays an important role in SMP. It comes under the category of smart materials whose physical properties vary according to external stimulation. The external stimulation may be pH, moisture, temperature, voltage/current, magnetism, etc. Upon this stimulation, the material responds and gains the required shape and size, and once this stimulation is removed, it will return to its original shape (Liu, Qin, and Mather 2007; Tiwari and Shaikh 2022b). Nature has always served as a source of inspiration for the development of numerous stimulus-responsive systems. Engineers have tried to copy nature and come up with materials and methods that would react to environmental conditions in the same way that nature does. The camouflaging of a chameleon or zebrafish, a squid changing its body color to match its surroundings, and the *Mimosa pudica* (touch-me-not) plant responding to the sensation of touch (Figure 1.1) are a few among many known examples.

Presently, technology has adapted to this 'smart' material, trying to push into regular working to reduce overall weight and volume occupied by associated machine components. It is very clear that the incorporation of new smart materials into traditional machines and mechanisms will reduce overall cost, weight, and volume occupied. Moreover, with this kind of new materials, complications associated with mechanism operation may be minimized, which can make systems more effective, versatile, and simple (Behl and Lendlein 2007; Tiwari and Shaikh 2022d).

Normal
Condition

Release of Moisture
due to Touch

FIGURE 1.1 Stimuli response from the nature of the *Mimosa pudica* plant leaf.

DOI: 10.1201/9781003391760-1

With minimal use of mechanisms in machine elements, friction and wear will also be reduced. It is a well-known fact that in a world of total energy consumption, about 80% is consumed to overcome friction and wear.

1.1   DISCOVERY OF SHAPE MEMORY EFFECT

Arne (Ölander 1932) first observed the shape memory effect (SME) while working with gold-cadmium alloy. Further, this term 'SME' was properly designated by Vernon and Vernon (1941) for his polymeric dental material. The SME in a nickel-titanium alloy was discovered by Buehler, Gilfrich, and Wiley (1963). The alloy is also known as nitinol, which is an amalgamation of the elements Nickel (Ni) and Titanium (Ti). From that point forward, the interest for shape memory alloys (SMAs) for designing and specialized applications has been expanding in various business fields, for example, in customized items and modern applications, construction, composites, cars, aviation, small actuators, flexible electronics, and biomedical (Wang, Liu, and Leng 2016; Tiwari and Shaikh 2022c).

In the early 1960s, a Japanese researcher named Yoshio Tsujimoto discovered the SME in a polymer, which he called "Shape Memory Plastic". However, it wasn't until the 1980s that researchers began to develop SMPs for practical applications. In recent times, shape memory materials and designs certainly stand out due to their recognized multifunctional properties and wide applications in aviation, vehicles, missiles, clinical gadgets, data capacity, energy collection, etc. It is a challenge to completely comprehend the complex multifunctional practices of different smart materials and constructions, and designing is important to improve the presentation and unwavering quality of these materials and constructions in modern applications (Mahinroosta et al. 2018).

1.2   CLASSIFICATION OF SHAPE MEMORY MATERIALS

SME can be generated due to numerous stimuli—be it moisture, heat, light, magnetic field, electricity, or chemicals. The effects are named after their change—inducing stimuli such as hydroactivity, electroactivity, thermoactivity, or magnetoactivity. Figure 1.2 indicates classification of the shape memory materials (SMMs) based on different stimulus responses. Among the various depicted stimuli, this literature will primarily focus on the temperature-stimulated smart materials. SMMs can be classified into four main categories based on their composition and properties:

1. **Metal-based SMMs:** These materials are made of metals and their alloys such as copper, nickel, titanium, and iron. They have a high SME, excellent mechanical properties, and good biocompatibility, which make them suitable for various applications in industries like aerospace, automotive, and biomedical.
2. **Polymer-based SMMs:** These materials are made of polymers (such as polyurethane, polycaprolactone, and polyethylene) with shape memory properties. They have lower mechanical strength compared to metal-based SMMs, but are lightweight, biocompatible, and can be processed easily. Polymer-based SMMs are used in biomedical applications such as tissue engineering, drug delivery, and medical implants.

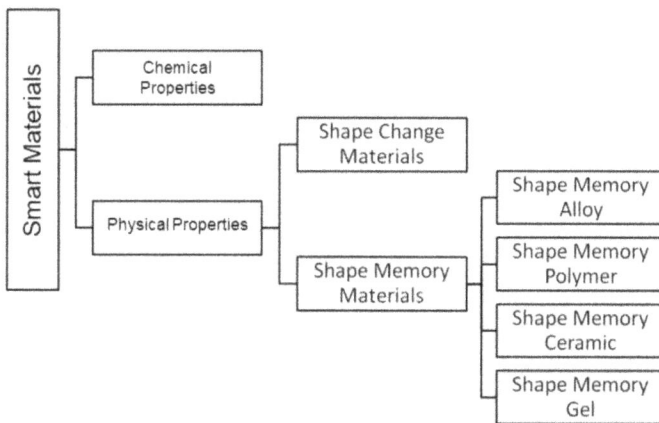

FIGURE 1.2   Classification of smart/shape memory materials.

3. **Ceramic-based SMMs:** These materials are made of ceramics with shape memory properties, such as zirconia and alumina. They have high hardness, excellent wear resistance, and good biocompatibility, which make them suitable for applications such as cutting tools and biomedical implants.
4. **Composite SMMs:** These materials are made of a combination of two or more materials, such as metal matrix composites and polymer matrix composites. Composite SMMs have superior properties compared to their constituent materials, such as high strength, stiffness, and SME. They are used in various applications such as aerospace, automotive, and biomedical.

Overall, the choice of SMM depends on the specific application requirements such as mechanical strength, biocompatibility, and processing methods. This work explains the fabrication techniques, characterization, micro-mechanism, and finite element analysis of polymer-based SMMs and their composites.

1.3   SHAPE MEMORY EFFECT IN POLYMERS

SMPs are materials that can be controlled to retain a deformed shape, which is then fixed under appropriate constraints. The change is regulated by temperature changes across the glass transition temperature of the polymer. The morphology and structure in the transition region are responsible for the SME of the polymer (Tiwari and Shaikh 2022a, 2021b). The shape memory cycle involves deforming the polymer into an intermediate shape through heating above its glass transition temperature, cooling and curing the material into an intermediate orientation, and then elevating the temperature above the transition temperature to retrieve the original shape. SMPs offer mechanical action triggered by an external stimulus and can be used for a range of applications due to their unique properties.

Figure 1.3 illustrates the entire process of a SME for the temperature-sensitive SMP. The process initiates when the temperature of the polymer is raised to glass

Original Shape   Programmed   Recovered
                            Shape   Shape

              Heating   **Cooling**   **Heating**

$T < T_{Trans}$ $T > T_{Trans}$ $T < T_{Trans}$ $T > T_{Trans}$

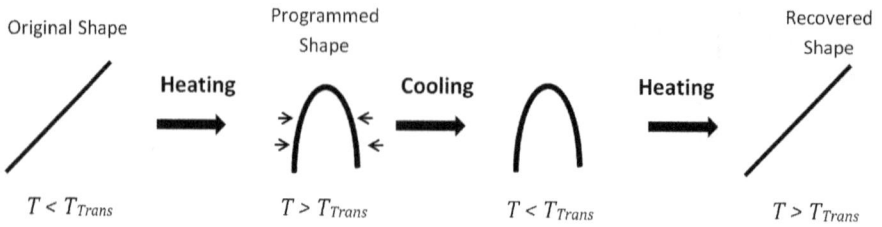

FIGURE 1.3   Flow chart of shape memory cycle (Leng et al. 2011).

transition temperature (T_{trans}), the temperature at which glassy substances become soft and rubbery. The polymer deforms under the application of external stress at an elevated temperature. Then the polymer is cooled to the operating temperature. The intermediate orientation is fixed, and stress or external loading can be removed, but it remains intact in the deformed shape. When it needs to release the strain and return to its original shape, the polymer will once again be heated to its transition temperature (T_{trans}).

1.4   GOVERNING PRINCIPLES OF SHAPE MEMORY EFFECT IN POLYMERS

SMPs have the ability to "remember" one or more shapes that are determined by their network elasticity. These shapes can be temporarily altered by immobilizing the material through vitrification or crystallization. For example, a complex three-dimensional SMP can be compressed into a slender form that can be easily delivered to the body or fit into a small space by heating, deforming, cooling, and unloading. Later, when exposed to heat, light, or solvent, the material will return to its original complex shape through network chain mobilization. While it may seem that many polymers could exhibit this SME, it actually involves a great deal of polymer chemistry and physics to fully understand and control (Tiwari and Shaikh 2019, 2021a). The ability to manipulate the properties of SMPs is a testament to researchers understanding of structure-property relationships in polymers.

The field of SMP research has seen significant progress in recent years, with many innovations and developments that highlight the unique properties of these materials. The evolution of stress, strain, and temperature during thermomechanical cycling of an SMP has been described to understand the mechanism of SMPs. The shape memory cycle can be visualized using the example of crosslinked poly(cyclooctene), a semi-crystalline network polymer, as shown in Figure 1.4. The curve starts from point (iv), where the temperature of the polymer is brought to T_g (the temperature where glassy substances become soft and rubbery). The rising portion along the stress axis denotes the shape-training process from point (iv) to point (i). The material is given its temporary shape at point (i) and is then cooled along the line (i)–(ii). On reaching point (ii), the intermediate orientation is constrained, and stress/external loading could be removed, returning along the stress axis through the line (ii)–(iii). This is the operating temperature of the polymer. The polymer will again be heated to its glass transition temperature at point (iv) (here, T_g) when required to release the strain and return to its original shape.

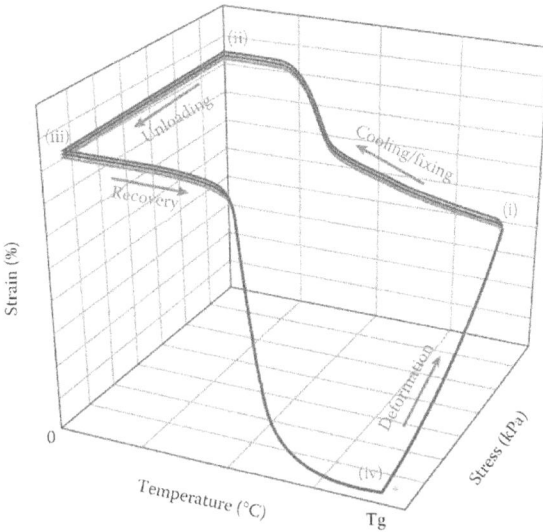

FIGURE 1.4 Shape memory polymers (SMPs) like crosslinked poly(cyclooctene) undergo thermomechanical cycles (Kunzelman et al. 2008).

SMPs are unlike most polymeric materials because they can hold a temporary shape and return to their original shape with the help of external forces (like heat, an electric field, a magnetic field, or radiation). Shape fixing and shape recovery are important because they show how the microstructure changes and how much SMPs can be used in the real world. Therefore, they are both fundamental and useful. Clearly, there are many different magnitudes of merit that can be thought of, such as the quality of fixing and recovery, the sharpness of the recovery process, and the work capacity, among others.

SMPs are sensitive to external stimuli such as heat, an electric field, a magnetic field, or irradiation. This ability sets them apart from conventional polymeric materials and is of both fundamental and practical importance. The features of fixing, recovery, and sharpness of the recovery event can be quantified to determine the suitability of SMP for practical applications.

Shape fixing, also known as fixity, involves the characteristics of retaining a temporary state by cooling below a transformation temperature, thereby storing strain energy. The measure of shape fixing or fixity, $R_f(\%)$, is determined by the ratio of the strain after unloading (ε_u) to the temporal strain achieved after deformation (ε_m). It may depend on the shape memory cycle number, N, as indicated in Eq. 1.1. At the molecular level, shape fixing can be achieved by designing the constituent chains to crystallize or vitrify at a specific temperature or immobilizing the chains using a secondary and labile crosslinked network.

$$R_f(\%) = \frac{\varepsilon_u}{\varepsilon_m} \times 100 \qquad (1.1)$$

By delivering a certain stimulus, it is possible for an SMP to revert to its "memorized" form, which is its permanent shape, after it has been temporarily fixed. In the research on SMPs, the recovery, also known as the change in stress and strain over the course of time and temperature, is of the utmost significance. Free recovery, also known as recovery, when there is no mechanical strain, is the approach that is used the most often in order to characterize free recovery. In this approach, a variety of recovery methods have been described. The shape recovery (R_r) is the measure that is used the majority of the time, and its equation can be found in Eq. 1.2 as

$$R_r(\%) = \frac{\varepsilon_u - \varepsilon_p}{\varepsilon_m - \varepsilon_p} \times 100 \tag{1.2}$$

where ε_u, ε_p, and ε_m represent the fixed strain after unloading, the permanent strain after heat-induced recovery, and the temporal strain achieved by deformation, respectively.

1.5 MOLECULAR MECHANISM OF THE THERMALLY INDUCED SHAPE MEMORY EFFECT

SMPs consist of networks of elastic polymers that have switches that respond to certain stimuli. The molecular switches and netpoints make up the polymer network (Figure 1.5). The netpoints impart the polymer network its permanent shape. They can be chemical (like covalent bonds) or physical (like interactions between molecules). Physical cross-linking happens in a polymer whose structure is made of at least two separate domains, like block copolymers. Here, chain segments in domains with the second highest thermal transition temperature (T_{trans}) act as molecular switches, while chain segments in domains with the highest thermal transition temperature (T_{perm}) act as netpoints (hard segments). If the working temperature is higher than T_{trans}, the switching domains are flexible. This gives the polymer network above T_{trans} an entropic elastic behavior. If an outside force has changed the shape of the sample before,

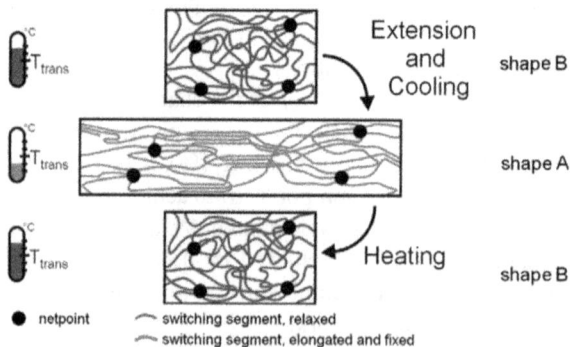

FIGURE 1.5 Molecular mechanism of the thermally induced shape memory effect T_{trans} = thermal transition temperature related to the switching phase (Lendlein and Kelch 2002).

when that force is removed, the sample snaps back into its original shape. Figure 1.5 shows how the SME works at the molecular level for the SME caused by heat.

The SMP network is made up of covalent netpoints and segments that switch based on how they interact physically. For shape memory to work, the polymer network has to be temporarily fixed in a deformed state under conditions that are right for the application. This is done by using reversible netpoints as molecular switches to stop the deformed chain segments that are under external stress from returning to their original shape. The extra netpoints could be attained from physical interactions or from covalent bonds. Physical cross-linking happens when T_{trans} related domains harden or crystallize. These switching domains can be made either by the chain segments themselves, which drive the entropic elastic behavior, or by the side chains, whose joining together can temporarily stop the side chains or side chain segments from recoiling. Attaching functional groups to the chain segments leads to covalent cross-linking that can be undone. These functional groups must be able to form covalent bonds that can be broken by reacting with each other or with functional groups that are a good match. Most of the SMPs that have been talked about so far respond to heat. In this case, the extra crosslinks are broken apart by heat, which causes the stimulation. If these extra crosslinks are caused by physical interactions, T_{trans} can be further divided into a glass transition T_g and a melting transition T_m. While glass transitions can happen over a wide range of temperatures, melting happens over a small range of temperatures. If the crosslinks are functional groups that can go through photoreversible reactions, the shape memory technology can be expanded to use light as a stimulus. Indirect ways to heat the material include using electrical currents or electromagnetic fields.

1.5.1 TRANSITION TEMPERATURE

Phase changes are achieved by training the polymeric material at a specific high temperature, defined as the transition temperature (T_{trans}) or glass transition temperature (T_g). The T_g of a polymer represent the range of temperatures at which mechanical properties are volatile. It is constantly lower as compared to the melting temperature (T_m) of the crystalline state of the material, if that exists. Figure 1.6 illustrates the drastic change in the storage modulus in the vicinity of T_g. The glass transition temperatures of various polymers and their composites are discussed in the respective sections on various polymers.

1.6 THE ROLE OF CRYSTALLIZATION

The application of external stress is the initial stage in the programming of a traditional one-way shape memory effect (OWSME) in a polymeric SMM. When the external stress is removed, the polymer returns to its original shape because chemical or physical modifications have been triggered to prevent further deformation. The material maintains its temporary form until an external stimulus is applied that overcomes the material's internal structural barrier and causes the shape to change. Thus, the stimulus sets in motion what is known as the OWSME (Figure 1.7), whereby the original form is recovered. It is crucial for the commercialization of SMMs that

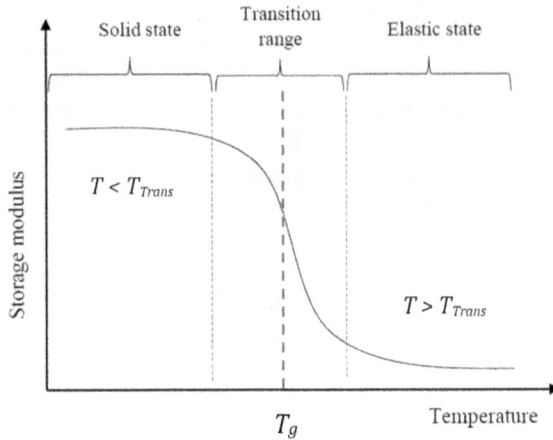

FIGURE 1.6 T_g expressed as a range of temperatures between the solid and elastic states.

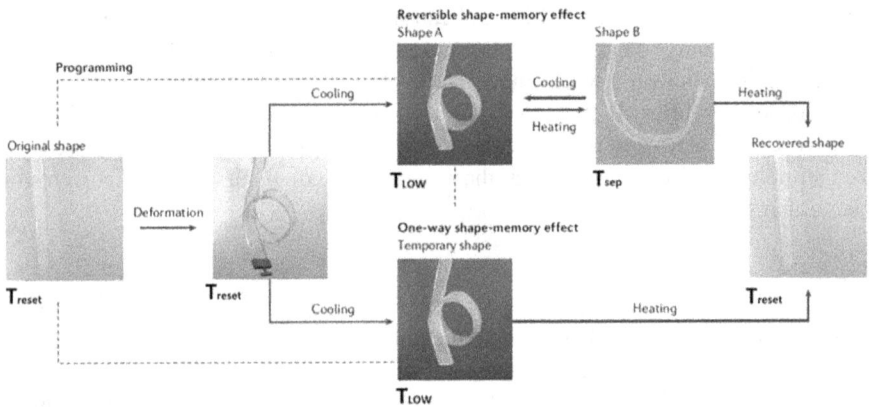

FIGURE 1.7 Programming of a coordinated physical response in polymeric systems (Lendlein and Gould 2019).

different stimuli may be employed for the programming and recovery processes, and several of these can coexist in the same multisensitive material.

When an external temperature rise causes a phase shift, such as the melting enthalpy of crystalline switching domains, the result is the OWSME. This dissipates potential entropy and results in useful mechanical work. Crystallization and melting of the actuation units of SMP actuators directly contribute to reversible motion. Following melting, entropy increases, contributing to shrinkage. When the chains are just partly aligned, however, they become immobile. By compressing and expanding a set of rubber spokes, the entropy wheel converts a temperature difference into mechanical energy. Without crystallizable actuation domains, macroscale deformation may be driven by the coordinated molecular reorientation of an amorphous domain. The entropy wheel's actuators are driven exclusively by entropic change, with continuous external stress standing in for geometry-determining domains. For a shape memory actuator that can be used in both directions, this raises the issue of whether crystallization is required.

Numerous liquid crystalline elastomer materials can generate sufficient energetic gain through macromolecular entropic changes without the need for crystallization and melting, as demonstrated by the entropy wheel. However, coordinated network deformation relies on volume change being directed by orientation. The entropy wheel employs external stress, which is neither programmable nor self-contained inside the same material. Instead of crystallizable geometry-determining domains, programmable and reprogrammable deformation might be achieved using molecular switches, which would provide the network anisotropy required to direct the alignment of the actuation domains.

1.7 SHAPE MEMORY CHARACTERIZATION

To measure the SME, mechanical testing can be done on a larger scale. The shape recovery ratio (R_r) and shape fixity ratio (R_f) are crucial values to determine for one-way SMMs. R_r measures how successful the process of recuperation is by comparing the original shape to the recovered shape. R_f measures the programming efficiency by comparing the fixed deformation ε_u to the total deformation ε_m. Cyclic thermomechanical testing is commonly used to determine these values. Controlling stress, strain, and temperature over the configuration and recovery processes is possible using a conventional mechanical testing setup that includes a thermos chamber. Compression, bending, and tensile testing are the most common techniques for material deformation. These methods may be supplemented with more complicated procedures such as shear deformation and torsion to offer a thorough thermomechanical description of a material.

The simplicity and one-directional nature of tensile testing have made it a widely used method for modeling the shape memory process. Cyclic thermomechanical testing results are often presented as stress–temperature–strain graphs that show the mechanical behavior at each stage of the shape memory cycle (Figure 1.8). The sample is heated to the deformation temperature T_{trans}, and an external force is applied,

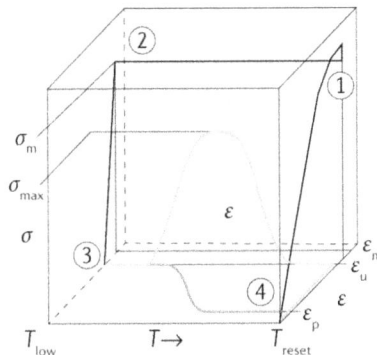

FIGURE 1.8 Three-dimensional representation of the one-way shape memory effect (Sauter et al. 2013).

causing an increase in stress until the strain reaches ε_m. The sample is then equilibrated at constant strain for a specific holding time to allow for initial stress relaxation.

After that, a stress-controlled procedure is initiated, in which the stress at the total deformation σ_m is kept constant even as the material cools. Finally, the strain reduces once the external force is withdrawn, and this gives us a measure of the sample's fixed strain, denoted by ε_u. After being programmed, the sample is recovered to the recovery strain ε_r by heating it over T_g. SMMs may be tested for reliability by plotting strain against time over numerous cycles to evaluate motion resilience and establish the time dependence of the shape change (Figure 1.9). Any variation in strain from cycle to cycle would be readily apparent here. Both stress-free and constant-strain situations (Figures 1.8 and 1.9) are suitable for the healing process. In a stress-free recovery procedure, the strain of the material is permitted to alter throughout the shape recovery, while in a constant-strain situation, the specimen's geometry determines the range of motion. Reorientation of macromolecules along chain segments generates a force that acts as a recovery stress on the device.

Stress-free recovery testing helps determine the temperature with the highest recovery rate, referred to as the swapping temperature T_{sw}, which is the tipping point on the stress–temperature curve (Figure 1.10). Additionally, stress-free recovery testing provides characteristic values, including the recovery temperature range ΔT_{rec} and the recovery rate v_{rec}.

On the other hand, the inflection point of the stress–temperature curve, $T_{\sigma,inf}$, and the maximum recovery stress temperature, $T_{\sigma,max}$, may be calculated during a constant-strain recovery process (Figure 1.11). The data gathered from these two types of recoveries is mutually beneficial. Stress-free recovery in polymeric SMMs is similar across physically crosslinked thermoplastics and a covalently crosslinked polymer network. However, when subjected to a steady tension, the materials act quite differently. The drop in Young's modulus of the pliable material is canceled out by the entropic elasticity of the thermoplastic sample at $T_{\sigma,max}$.

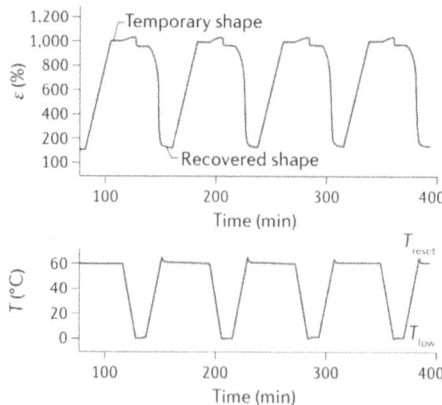

FIGURE 1.9 Illustration of cyclic testing (Mazurek-Budzyńska et al. 2017).

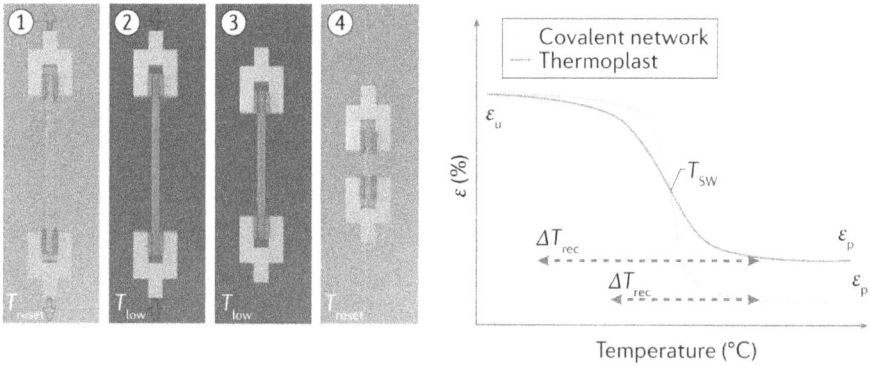

FIGURE 1.10 Approach for programming diagrams that allow for stress-free recovery (Sauter et al. 2013).

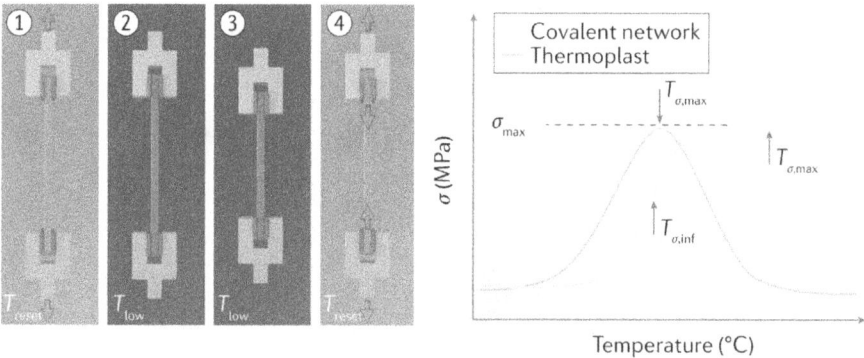

FIGURE 1.11 Schematic illustration of shape memory creation procedure with constant-strain recovery (Lendlein and Gould 2019).

On the other hand, outside the scope of physical net points, the rising chain mobility of physical net points $T_{\sigma,\max}$ in the covalently crosslinked polymer network, leads to a significant decrease in stress (Lendlein and Gould 2019).

A reversible shape memory actuator undergoes a programming process involving deformation and fixation steps in a manner similar to that of unilateral shape memory (Figure 1.12). Deformation is carried out at a temperature T_{reset} higher than both crystalline domains' melting points, and fixation is achieved by cooling to a temperature T_{low} below the melting transitions. A reversible change in the sample's shape is produced by heating and cooling between a temperature T_{sep}, the middle point between the two melting phases, and T_{low}. To characterize the reliability of the actuation behavior of a reversible two-way SME, multicyclic measurements are essential (Figure 1.13). During multicyclic testing, no external force is applied, and unlike with an OWSME, numerous actuation cycles may be created from a single programming operation. A strain-temperature plot (Figure 1.14) shows the morphological transition between shapes A and B that occurs throughout the actuation cycle

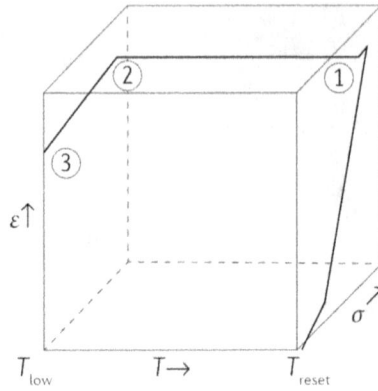

FIGURE 1.12 Thre-dimensional representation of a reversible shape memory effect (Lendlein and Gould 2019).

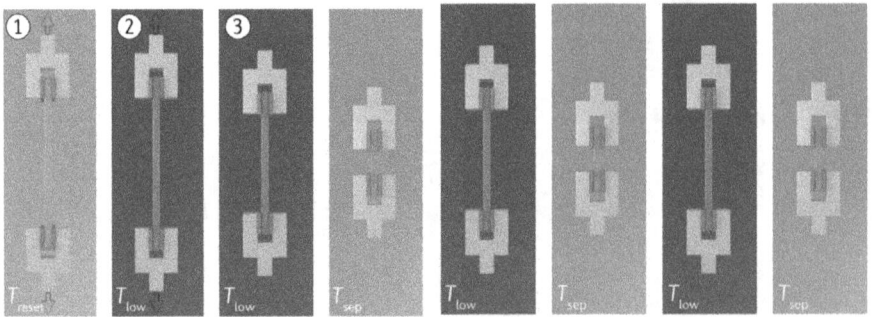

FIGURE 1.13 Schematic illustration of programming with continuous actuation (Lendlein and Gould 2019).

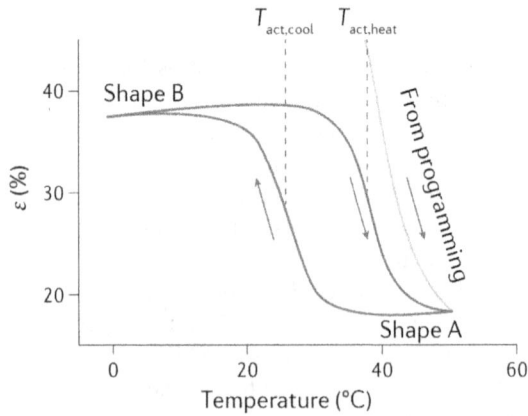

FIGURE 1.14 Schematic illustration of curve showing the actuation of a sample under stress-free conditions (Lendlein and Gould 2019).

of a reversible SMP actuator. Additionally, the greatest heating and cooling actuation rates may be identified with the help of this graph. Even though stress-free measuring methods may be employed to examine reversible memory effects in free-standing objects, it is possible to elicit a more significant shape change by introducing external stress during the actuation cycle (Figure 1.15).

1.7.1 MORPHOLOGY-DEPENDENT FUNCTION

SMMs' physical characteristics significantly influence both their functioning and performance, while adjusting device shape may generate extra hierarchical force. Different morphologies have been studied to meet diverse application requirements. Although the basic molecular processes of shape memory that are responsible for each of these morphologies are the same, the macroscale shape and the microscale structure may be modified to raise the actuation magnitude and introduce unique behaviors such as bending, twisting, stretching, and folding. In nature, certain morphological changes, such as the folding of bird wings and flowers, can enhance actuation. However, in the case of SMPs, engineering principles are typically borrowed from the mechanical engineering of machines. These machines have shape memory metallic alloys that are limited to a 3%–4% reversible change in strain. However, the incorporation of a folding mechanism into SMPs has greatly enhanced their amplitude of actuation. This is particularly the case for a two-way reversible SME, in which the amplitude of actuation is typically confined to length variations of up to 20%.

A ribbon made of poly[ethylene-co-(vinyl acetate)] can undergo length variations that may be reversed to about 100% by being programmed into a concertina shape (as shown in Figure 1.16). Similarly, folding techniques found in nature can be adapted to build three-dimensional objects out of polymer sheets that respond

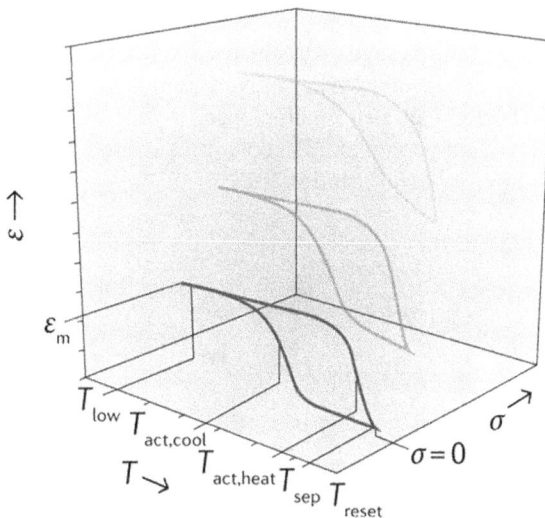

FIGURE 1.15 Example of recovery under different applied stresses (Lendlein and Gould 2019).

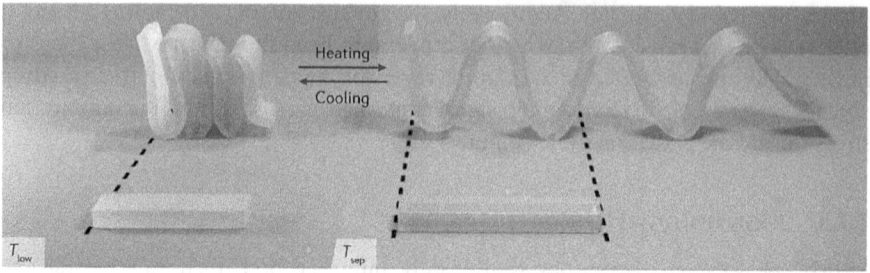

FIGURE 1.16 Zig-zag pattern of shape memory polymers (Lendlein and Gould 2019).

to external stimuli such as light, heat, magnetic fields, and solvent. However, there are still obstacles to overcome in managing the orientation of chemical chains in layer processing techniques and forecasting changes in multidimensional structure. Using thermal actuation induced by electromagnetic fields, it has been possible to make SMMs fold themselves under various stimuli, including light, Joule heating, and thermal radiation. By incorporating stimuli-responsive elements and origami-inspired designs, multifunctional soft robots can be created, such as a temporary 3D crane shape from a polymer sheet, which utilizes a thermoresponsive structural unit and light as a spatioselective stimulus to enable flapping movements of the wings (as shown in Figure 1.17).

Using as much light as is possible to pre-program a spatioselective stimulus, which is an active zone inside the polymer film, to enable the wings' flapping movements. The crane can more easily perform the needed reversible actuation movement because the linear elongation-contraction movement has been localized to a prefabricated joint. This has resulted in the crane exhibiting a '4D' behavior. Combining a conventional SME with a two-way SMP actuation that acts at a larger temperature range or by the production of extra crosslinks is what makes this behavior conceivable. Alternative techniques such as fused deposition modeling or UV curing can also be employed to generate 2D structures, which are then transformed into complex 3D geometries using SMEs, thereby avoiding time-consuming layer-by-layer printing.

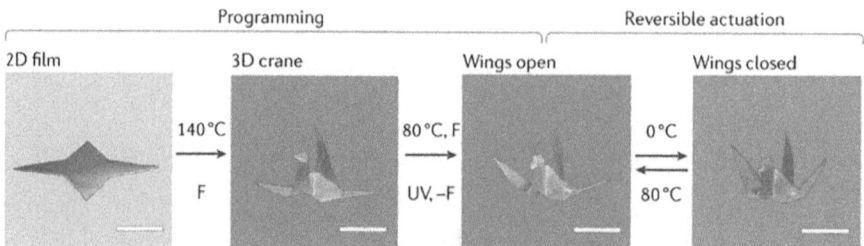

FIGURE 1.17 Shape memory illustration in 2D and 3D orientation (Lendlein and Gould 2019).

1.8 APPLICATION

The adaptability of the shape memory process lends itself well to the development of practical applications for nonisotropic and irregularly shaped transformations. For instance, the radial shrinking of heat-shrink tubing may be performed by the shape memory process, which is something that cannot be accomplished using other technologies such as liquid crystalline elastomers. They can be programmed using proven polymer processing techniques that are scalable for industrial usage, and their simplicity makes the polymers that can be used in SMP-based goods financially feasible. SMP-based products have the potential to revolutionize a wide range of industries. As a result of the presence of soft materials inside shape memory devices, which have mechanical characteristics that are analogous to those of soft tissues, these devices are well suited for a variety of applications in the biomedical industry. In order to regulate and control the flow of fluids via intricate networks of micrometer-sized channels, microvalves, which are often utilized in microfluidic systems for biochemical research, are typically required. Without the need for an additional control system, SMMs may serve as the basis for an actuation mechanism that is both dependable and resilient. For instance, the recovery of a microvalve that has been activated by heat may be used to link or disconnect two neighboring microchannels. This eliminates the need for cumbersome pneumatic supply systems and enables the design of devices that are more compact. It is possible to increase the thermally responsive polymer's biological compatibility by modifying the amount of branching and the molecular mass of the polymer network in order to control the material's melt transition temperature ($T_m = 60°C$). In addition, a valveless microfluidic control mechanism may be created by programming the whole channel and making use of local heating to cause a spatially focused shape change (Figure 1.18).

The use of microstructured channel walls can offer specific functionalities, such as protein absorption and selective wetting behavior, by providing control over adhesion and friction parameters. Additionally, surface nanopatterning can be utilized to provide signaling cues for adhesion-mediated cell signaling mechanisms. Anisotropic topographical features can facilitate the alignment and migration of various cell types, a process known as contact guidance. SMMs have enabled the creation of dynamically reconfigurable surface topologies that can be controlled by spatially localized and environmental stimuli. These dynamic cell culture platforms with shape memory can regulate cell function in response to cell-compatible stimuli.

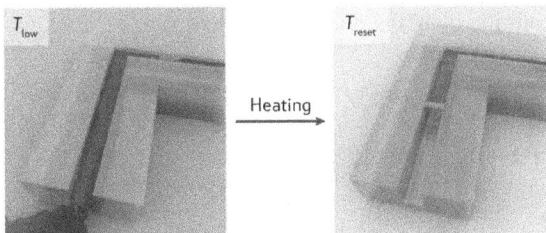

FIGURE 1.18 Temperature-regulated microfluidics channel (Ebara et al. 2013; Lendlein and Gould 2019).

For instance, incorporating gold nanorods into a polymer film allows near-infrared light to induce photothermal heating that triggers a spatially localized recovery. This topological change brought about by the OWSME has a direct impact on the morphology and migration of fibroblasts.

The need for clinical procedures that are less invasive has resulted in the utilization of shape memory technology for clinical devices. Small objects that unfold or expand into a desired shape after implantation can be utilized to enhance the opening of blocked vessels or wound closure. Moreover, the use of a biodegradable SMP suture, which is composed of a thermoplastic block copolymer, can provide a time-efficient approach for wound closure through heating (Figure 1.19). The benefits of SMP-based sutures are that they are biodegradable and offer a defined pressure on the wound, which is advantageous for the healing process. Additionally, one-way shape memory capability has been integrated with drug release or therapeutic activities, such as a mechanically active ureter stent with incorporated antibiotic activity. Such multifunctional devices that provide controlled drug release, are hydrolytically degradable, and exhibit shape memory may also be useful for tissue engineering applications.

The adaptability of the shape memory process lends itself well to the development of practical applications for nonisotropic and irregularly shaped transformations. For instance, the radial shrinking of heat-shrink tubing may be performed by the shape memory process, which is something that cannot be accomplished using other technologies such as liquid crystalline elastomers. They can be programmed using proven polymer processing techniques that are scalable for industrial usage, and their simplicity makes the polymers that can be used in SMP-based goods financially feasible. SMP-based products have the potential to revolutionize a wide range of industries. As a result of the presence of soft materials inside shape memory devices, which have mechanical characteristics that are analogous to those of soft tissues, these devices are well suited for a variety of applications in the biomedical industry.

Due to their ductility and environmental responsiveness, SMPs are great candidates for the development of self-regulating materials for smart textiles that react to changes in temperature. These materials might be used in temperature-responsive clothing. Commercial lines of athletic wear have begun using SMPs, which enables the creation of multifunctional garments that are not only windproof, waterproof, and breathable but also have the capacity to have temperature-dependent permeability to moisture. Fabrics that change form in response to temperature may also be made by combining SMP fibers with flexible and lightweight yarns. When heated to 50°, a polyurethane and nylon mix that has been flattened may be made to rebound to a shrunken condition by being programmed to do so.

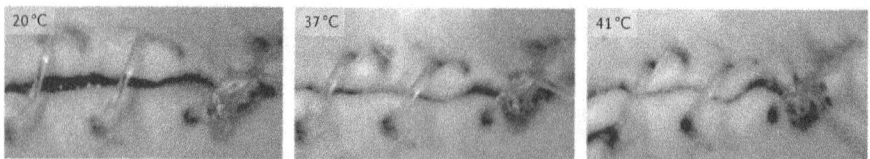

FIGURE 1.19 Biodegradable shape memory polymer suture (Lendlein and Langer 2002; Lendlein and Gould 2019).

Recent developments in SMAs and SMPs (SMPs) have motivated researchers to construct intelligent textiles as self-regulating, shape-changing structures that adapt to environmental variation. This has contributed to a new area in the scientific frontier of smart materials and created a new field in the scientific frontier of smart materials. This research demonstrates how engineering fabric designs with adequate shape memory training may improve both the functionality and aesthetic appeal of the finished product. The training technique of SME in SMAs and SMPs, the manipulation of phase transition temperatures in SMAs, the fabrication of SMP yarns, and SMP coating on textiles for particular needs of fabric design are all considered to be primary technologies. For the purpose of achieving certain dynamic aesthetic effects, more consideration is also given to the practice of combining SMPs or SMAs with the textile design of various types of fabric yarn.

The conventional actuators used in robotics are often cumbersome and rigid, which restricts the range of possible uses for these devices. On the other hand, components that are soft, flexible, and constructed of lightweight materials provide an option that is more adaptable. In particular, reversible SMP actuators give a chance to overcome the constraints of one-way shape memory and replace the requirement for controlling units that are necessary in pneumatic systems. This may be accomplished by overcoming the restrictions of one-way shape memory. As an illustration of this, consider the design of a twisted, non-continuously responding actuator (Figure 1.20), which can flip an arrow sign between three distinct positions and demonstrates the potential for behavioral complexity and geometric variety. SMP actuation technologies will be able to profit from the capacity to produce complicated and high-performing components as 3D printing and multimaterial fabrication processes continue to advance. For example, stereolithography has been utilized to construct SMP microgrippers (Figure 1.21), which are able to take up and release things in

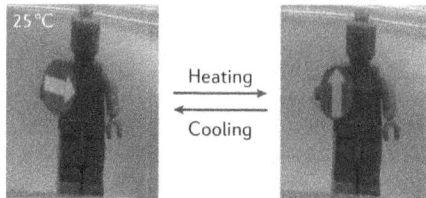

FIGURE 1.20   Shape memory polymer actuation in robotics (Farhan et al. 2017).

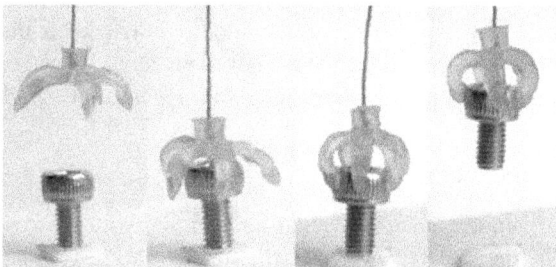

FIGURE 1.21   Shape memory microgripper (Ge et al. 2016).

response to variations in temperature. The addition of alternative materials at the tips of the grippers, such as polymethylmethacrylate, may further permit modification in stiffness for the purpose of maintaining safe contact with a variety of items.

1.9   OUTLOOK CHALLENGES

It is required to integrate macromolecular architecture with macroscale processing in order to provide programmable recovery and actuation in SMPs. This can only be done by combining the two concepts. The scientific community still needs to focus on improving the spatial localization of the material response in order to increase the complexity of SMP actuators. Although there are several methods available for achieving an OWSME and reversible SMP actuation, neither of these effects can be reversed. This may be accomplished by inventing processing techniques that allow for the changing of local network architecture, such as altering the distribution of cross-links within a polymeric material. Another way to accomplish this goal is by managing the distribution of atoms inside a material. In addition, the creation of biomimetic SMP actuation-based multifunctional devices is made possible by multimaterial systems such as hybrids of polymers and gels. In particular for biomedical devices and applications that call for a localized reaction, it would be very intriguing to create SMP actuators that react to stimuli other than heat. In the field of robotics, having the capability to initiate morphological changes in response to biological inputs might have considerable benefits over currently used design and production methods.

Even though the invention of SMMs has led to the production of a wide variety of stimuli-responsive polymeric systems, there are still challenges to be overcome in order to achieve multifunctionality. This is especially true for the sequential coupling of functions such as heat generation and thermoresponse. It is essential to properly utilize the structural knowledge of SMPs in order to produce certain behaviors that are favorable to roboticists and design engineers in order for shape memory actuation to be a success. This will allow for specific behaviors to be achieved. The development of novel materials and technologies that can meet the needs of many industries, including information technology, soft robotics, and healthcare, will be facilitated through collaborations across diverse sectors. There is still a need for soft materials that can match the durability, accuracy, and reaction magnitude of hard pneumatic actuators, despite the fact that advanced manufacturing processes like 3D printing and soft lithography have broadened the variety of morphologies that are conceivable for shape memory devices. It is possible to accelerate the commercialization of SMP actuation technology by using low-cost, commodity polymers as the starting material for SMP actuators. This will lead to the development of a new generation of soft, reprogrammable actuators that are well suited to solve a variety of global concerns.

1.10   SHAPE MEMORY ALLOY

It is required to integrate macromolecular architecture with macroscale processing to provide programmable recovery and actuation in SMPs. This can only be done by combining the two concepts. The scientific community still needs to focus on improving the spatial localization of the material response to increase the complexity of SMP actuators. Although there are several methods available for achieving an OWSME and reversible SMP actuation, neither of these effects can be reversed. This may be accomplished by inventing processing techniques that allow for the changing of local network architecture, such as altering the distribution of crosslinks within a polymeric material. Another way to accomplish this goal is by managing the distribution of atoms inside a material. In addition, the creation of biomimetic SMP actuation-based multifunctional devices is made possible by multimaterial systems such as hybrids of polymers and gels. In particular for biomedical devices and applications that call for a localized reaction, it would be very intriguing to create SMP actuators that react to stimuli other than heat. In the field of robotics, having the capability to initiate morphological changes in response to biological inputs might have considerable benefits over currently used design and production methods.

SMAs are metallic alloys that show the SME, which enables them to restore their former shape after being distorted when they are subjected to specific external stimuli such as heat or stress. This effect gives them the ability to be classified as SMAs. Because of their capacity to alter their form in response to the action of external stimuli, SMAs are often referred to as "smart materials."

Nitinol, a nickel-titanium alloy, was developed in the 1960s by researchers at the Naval Ordnance Laboratory in White Oak, Maryland. Today, it is the most widely used kind of SMA. Because of the exceptional qualities that it has, such as high strength, low weight, and biocompatibility, nitinol is used in a broad number of different applications.

SMAs have a wide range of uses and may be found in places such as consumer electronics, medical equipment, and components for the aerospace industry. The following are some examples of goods that are based on SMA:

1. **Stents:** SMAs are used in the construction of stents, which are medical devices used to prop open blocked or narrow arteries in the body. Nitinol stents have the ability to expand and contract to the correct shape and size once they are implanted in the body.
2. **Orthodontic wires:** SMAs are used in orthodontic wires because they can be shaped into the desired shape and then return to their original shape once they are heated. This allows for more comfortable and effective orthodontic treatment.
3. **Actuators:** SMAs are used in a variety of actuators, which are devices that convert energy into motion. SMA actuators are used in aerospace components such as flaps and landing gear, as well as in consumer electronics such as cameras and mobile phones.
4. **Robotics:** SMAs are also used in robotics to create soft and flexible robots that can move and bend like natural muscles.

Overall, SMAs are versatile and unique materials that have a wide range of practical applications. They continue to be an active area of research and development for new and innovative applications.

In reality, SMAs are capable of existing in two distinct phases, having three distinct crystal structures (i.e. twinned martensite, detwinned martensite, and austenite), and undergoing a total of six different transformations. The structure of austenite is unaffected by temperature, but the structure of martensite is unaffected by temperatures below austenite's melting point. After being heated, an SMA will begin to change from the martensite phase into the austenite phase. The temperature at which this transformation begins is referred to as the austenite-start temperature (A_s), and the temperature at which this transformation is finished is referred to as the austenite-finish temperature (A_f). Once an SMA has been heated beyond its A_s, it will start to compress and change into the austenite structure, which means that it will revert back to its initial shape. This change is achievable even under large, applied loads, which results in high actuation energy densities as a direct consequence of this phenomenon. The transformation back to martensite begins at the temperature designated as the martensite-start temperature (M_s) and continues until it reaches the temperature designated as the martensite finish temperature (M_f) throughout the cooling phase.

The critical temperature is defined as the greatest temperature at which martensite can no longer be stress-induced. Above this temperature, the SMA is irreversibly deformed, just like any other metallic material. Figure 1.22 provides a concise summary of the SMA's shape memory behavior from an overall perspective. These effects, known as shape memory effects (SMEs) and pseudoelasticity

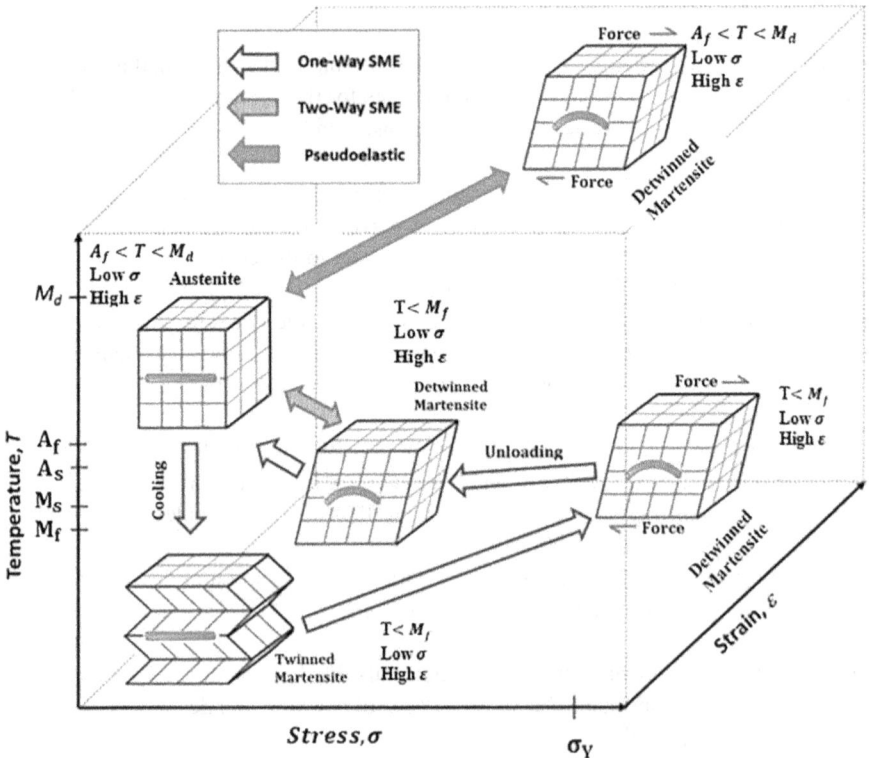

FIGURE 1.22 SMA phases and crystal structures (Mohd Jani et al. 2014).

(or superelasticity), may be broken down into three distinct categories of shape memory characteristics:

1. **One-way SME:** Following the removal of an external force, the one-way shape memory alloy (OWSMA) maintains its distorted condition until it is heated, at which point it returns to its initial shape.
2. **Two-way SME or reversible SME:** In addition to the one-way effect, a two-way shape memory alloy (TWSMA) may maintain its original form at temperatures that range from very cold to very hot. TWSMA, on the other hand, does not see as much use in the commercial sector because of the "training" requirements, the fact that it typically generates around half of the recovery strain that OWSMA does for the same material, and the fact that its strain has a tendency to disintegrate fast, particularly when exposed to high temperatures. As a result, OWSMA offers a solution that is both more dependable and more cost-effective. There are many other training techniques that have been offered, and two of these are spontaneous induction and load-assisted induction from the outside.
3. **Pseudoelasticity or Superelasticity:** The SMA returns to its initial form after being subjected to mechanical stress at temperatures ranging from A_f to M_d, and this occurs even in the absence of any thermal activation.

1.11 COMPARISON BETWEEN SHAPE MEMORY ALLOY AND SHAPE MEMORY POLYMER

SMAs and SMPs are two types of smart materials that exhibit the SME. While they share some similarities, there are also differences in their composition, properties, and applications.

- **Composition:** SMAs are metallic alloys, typically made from a combination of nickel and titanium, with some containing copper, iron, or cobalt. Nitinol, a nickel-titanium alloy, is one of the most commonly used SMAs. On the other hand, SMPs are polymers made from various materials, including thermoplastics, elastomers, and thermosets. These polymers can be tailored for specific applications through the selection of monomers and the addition of fillers or reinforcements (Lendlein and Kelch 2002).
- **Properties:** SMAs exhibit high strength and stiffness, making them suitable for applications requiring these properties, such as aerospace components and medical devices. SMAs also have a high damping capacity, which enables them to absorb and dissipate energy, making them suitable for use in vibration-damping applications. SMPs, on the other hand, are generally more flexible and compliant than SMAs, which makes them useful for applications such as biomedical implants and soft robotics. SMPs can also

be designed to have shape memory properties that can be activated by various stimuli, such as heat, light, or mechanical stress.

- **Applications:** SMAs have been used in various applications, including aerospace components, medical devices, and consumer electronics. SMAs are used in medical devices, such as stents, orthodontic wires, and cardiovascular devices, due to their biocompatibility and ability to be shaped into complex geometries. In aerospace, SMAs are used in applications such as wing flaps, landing gear, and engine components due to their high strength, stiffness, and damping capacity. SMPs have been used in biomedical implants, such as sutures and tissue scaffolds, due to their biocompatibility and ability to conform to complex shapes. SMPs have also been used in soft robotics, such as grippers and actuators, due to their flexibility and compliance.

In summary, SMAs and SMPs are two types of smart materials with distinct compositions, properties, and applications. The choice of material depends on the specific application requirements and the properties of each material. A comparison of the different characteristics of SMPs and SMAs is summarized in Table 1.1. NiTi, nickel–titanium.

TABLE 1.1
Analysis of the Similarities and Differences between Shape Memory Alloys and Shape Memory Polymers (Liu, Qin, and Mather 2007)

	Shape Memory Polymers	Shape Memory Alloys
Density (g/cm³)	0.9–1.1	6–8
Extent of deformation (%)	Up to 800%	<8%
Young's modulus at $T < T_{Trans}$ / GPa	0.01–3	83 (NiTi)
Young's modulus at $T > T_{Trans}$ / GPa	$(0.1–10) \times 10^{-3}$	28–41
Stress required for deformation/ MPa	1–3	50–200
Stress generated during recovery/MPa	1–3	150–300
Critical temperatures/°C	−10 to 100	−10 to 100
Transition breath/°C	10–50	5–30
Recovery speeds	<1 second to several minutes	<1 second
Thermal conductivity/Wm⁻¹ K⁻¹	0.15–0.30	18 (NiTi)
Biocompatibility and biodegradability	Can be biocompatible and/or biodegradable	Some are biocompatible (i.e. Nitinol), not biodegradable
Processing conditions	<200°C, low pressure	High temperature (>1,000°C) and high pressure required
Corrosion performance	Excellent	Excellent
Cost	<$10 per lb	~$250 per lb

REFERENCES

Behl, Marc, and Andreas Lendlein. 2007. "Shape-Memory Polymers." *Materials Today* 10 (4): 20–8. https://doi.org/10.1016/S1369-7021(07)70047-0.

Buehler, William J., John V. Gilfrich, and Robert C. Wiley. 1963. "Effect of Low-Temperature Phase Changes on the Mechanical Properties of Alloys near Composition TiNi." *Journal of Applied Physics* 34 (5): 1475–7. https://doi.org/10.1063/1.1729603.

Ebara, Mitsuhiro, Koichiro Uto, Naokazu Idota, John M. Hoffman, and Takao Aoyagi. 2013. "Rewritable and Shape-Memory Soft Matter with Dynamically Tunable Microchannel Geometry in a Biological Temperature Range." *Soft Matter* 9 (11): 3074–80. https://doi.org/10.1039/C3SM27243E.

Farhan, Muhammad, Tobias Rudolph, Ulrich Nöchel, Wan Yan, Karl Kratz, and Andreas Lendlein. 2017. "Noncontinuously Responding Polymeric Actuators." *ACS Applied Materials & Interfaces* 9 (39): 33559–64. https://doi.org/10.1021/acsami.7b11316.

Ge, Qi, Amir Hosein Sakhaei, Howon Lee, Conner K. Dunn, Nicholas X. Fang, and Martin L. Dunn. 2016. "Multimaterial 4D Printing with Tailorable Shape Memory Polymers." *Scientific Reports* 6 (1): 31110. https://doi.org/10.1038/srep31110.

Kunzelman, Jill, Taekwoong Chung, Patrick T. Mather, and Christoph Weder. 2008. "Shape Memory Polymers with Built-in Threshold Temperature Sensors." *Journal of Materials Chemistry* 18 (10): 1082–6. https://doi.org/10.1039/B718445J.

Lendlein, Andreas, and Oliver E. C. Gould. 2019. "Reprogrammable Recovery and Actuation Behaviour of Shape-Memory Polymers." *Nature Reviews Materials* 4 (2): 116–33. https://doi.org/10.1038/s41578-018-0078-8.

Lendlein, Andreas, and Steffen Kelch. 2002. "Shape-Memory Polymers." *Angewandte Chemie International Edition* 41 (12): 2034–57. https://doi.org/10.1002/1521-3773 (20020617)41:12<2034::AID-ANIE2034>3.0.CO;2–M.

Lendlein, Andreas, and Robert Langer. 2002. "Biodegradable, Elastic Shape-Memory Polymers for Potential Biomedical Applications." *Science* 296 (5573): 1673–6. https://doi.org/10.1126/science.1066102.

Leng, Jinsong, Xin Lan, Yanju Liu, and Shanyi Du. 2011. "Shape-Memory Polymers and Their Composites: Stimulus Methods and Applications." *Progress in Materials Science* 56 (7): 1077–135. https://doi.org/10.1016/j.pmatsci.2011.03.001.

Liu, Changdeng, Haihu Qin, and Patrick T. Mather. 2007. "Review of Progress in Shape-Memory Polymers." *Journal of Materials Chemistry* 17 (16): 1543–58. https://doi.org/10.1039/B615954K.

Mahinroosta, Mostafa, Zohreh Jomeh Farsangi, Ali Allahverdi, and Zahra Shakoori. 2018. "Hydrogels as Intelligent Materials: A Brief Review of Synthesis, Properties and Applications." *Materials Today Chemistry* 8 (June): 42–55. https://doi.org/10.1016/j.mtchem.2018.02.004.

Mazurek-Budzyńska, Magdalena, Muhammad Yasar Razzaq, Karolina Tomczyk, Gabriel Rokicki, Marc Behl, and Andreas Lendlein. 2017. "Poly(Carbonate-Urea-Urethane) Networks Exhibiting High-Strain Shape-Memory Effect." *Polymers for Advanced Technologies* 28 (10): 1285–93. https://doi.org/10.1002/pat.3948.

Mohd Jani, Jaronie, Martin Leary, Aleksandar Subic, and Mark A. Gibson. 2014. "A Review of Shape Memory Alloy Research, Applications and Opportunities." *Materials & Design (1980-2015)* 56 (April): 1078–113. https://doi.org/10.1016/j.matdes.2013.11.084.

Ölander, Arne. 1932. "An Electrochemical Investigation of Solid Cadmium-Gold Alloys." *Journal of the American Chemical Society* 54 (10): 3819–33. https://doi.org/10.1021/ja01349a004.

Sauter, Tilman, Matthias Heuchel, Karl Kratz, and Andreas Lendlein. 2013. "Quantifying the Shape-Memory Effect of Polymers by Cyclic Thermomechanical Tests." *Polymer Reviews* 53 (1): 6–40. https://doi.org/10.1080/15583724.2012.756519.

Tiwari, Nilesh, and AbdulHafiz A. Shaikh. 2019. "Flexural Analysis of Thermally Actuated Fiber Reinforced Shape Memory Polymer Composite." *Advances in Materials Research* 8 (4): 337–59. https://doi.org/10.12989/amr.2019.8.4.337.

Tiwari, Nilesh, and AbdulHafiz A. Shaikh. 2021a. "Micro Buckling of Carbon Fiber in Triple Shape Memory Polymer Composites under Bending in Glass Transition Regions." *Materials Today: Proceedings* 44 (6): 4744–8. https://doi.org/10.1016/j.matpr.2020.10.961.

Tiwari, Nilesh, and AbdulHafiz A. Shaikh. 2021b. "Micro-Buckling of Carbon Fibers in Shape Memory Polymer Composites under Bending in the Glass Transition Temperature Region." *Curved and Layered Structures* 8 (1): 96–108. https://doi.org/10.1515/cls-2021-0009.

Tiwari, Nilesh, and AbdulHafiz A. Shaikh. 2022a. "Effect of Size and Surface Area of Graphene Nanoplatelets on the Thermomechanical and Interfacial Properties of Shape Memory Multiscale Composites." *Polymer-Plastics Technology and Materials* 61 (12): 1–13. https://doi.org/10.1080/25740881.2022.2061864.

Tiwari, Nilesh, and AbdulHafiz A. Shaikh. 2022b. "Effect of the Amine and Carboxyl Functionalised Graphene on the Thermomechanical and Interfacial Properties of the Shape Memory Polymer Nanocomposites." *Polymer Bulletin*, 1–19.

Tiwari, Nilesh, and AbdulHafiz A. Shaikh. 2022c. "Effect of Graphene Nanoplatelets on the Thermomechanical Behaviour of Smart Polymer Nanocomposites." In *Aerospace and Associated Technology*, edited by Anup Ghosh, Kalyan Prasad Sinhamahapatra, Ratan Joarder, and Sikha Hota, 300–4. Routledge.

Tiwari, Nilesh, and AbdulHafiz A. Shaikh. 2022d. "Influence of Carboxyl-Functionalized Graphene Nanoplatelets on the Thermomechanical and Morphological Behavior of Shape Memory Nanocomposites." *Composites: Mechanics, Computations, Applications: An International Journal* 13 (3): 75–85. https://doi.org/10.1615/CompMechComputApplIntJ.2022043126.

Vernon, Lester B., and Harold M. Vernon. 1941. Process of manufacturing articles of thermoplastic synthetic resins. United States US2234993A, filed February 6, 1937, and issued March 18, 1941. https://patents.google.com/patent/US2234993A/en.

Wang, Wenxin, Yanju Liu, and Jinsong Leng. 2016. "Recent Developments in Shape Memory Polymer Nanocomposites: Actuation Methods and Mechanisms." *Coordination Chemistry Reviews* 320–321 (August): 38–52. https://doi.org/10.1016/j.ccr.2016.03.007.

2 Shape Memory Polymer Composites

Shape memory polymers (SMPs) have intrinsic disadvantages, such as a low heat transfer rate and low rigidity and strength, particularly at temperatures above the transition temperature, despite the fact that they have desirable and distinctive characteristics that can be used to progress technology in a variety of sectors. This restricts their use because the majority of uses, especially those in the aircraft sector, call for high actuating forces that are equivalent to high retrieval forces for the SMPs (Pilate et al. 2016). Depending on the size of the category of reinforcement, smart composites are referred to in the scientific community either as SMPNCs or as SMPCs. When compared to pure SMPs, these have reportedly shown increased healing loads (Fejős, Romhány, and Karger-Kocsis 2012). The polymers also acquire novel capabilities in comparison to the traditional SMPs due to modified elasticity, porosity, color, transparency, conductivity, and shape memory characteristics performance. Thermal and nonthermal shape memory control techniques are presented. Additionally, it has been shown that various SMPCs can become multifunctional, including those in aeronautics, biology, medical devices, flexible electronics, lenient robotics, shape memory arrays, and 4D printing.

2.1 COMPOSITES

A summary of composite materials is provided in this section to provide a basic understanding of them, which would be useful during the discussion of the SMPC. A composite is a body composed of two or more component substances with various physical and molecular characteristics. These materials are combined to create a hybrid substance with better qualities than the sum of its parts. A matrix and a support substance make up composites in general. The reinforcement material can be fibers, particles, or other types of fillers that are added to enhance the characteristics of the composite material, while the matrix material is usually a polymer, metal, or ceramic that provides a continuous phase. By regulating the type, quantity, and direction of the reinforcement material as well as the core material, the characteristics of composites can be customized. As a result, materials with specific characteristics such as high strength, rigidity, hardness, or thermal and electrical conductivity can be tailored. Based on the shape of their constituents, number of layers, direction of the fibers, length of the fibers, etc., composites can be classified into various categories. Classification of materials is depicted in Figure 2.1, along with their corresponding categorization.

DOI: 10.1201/9781003391760-2

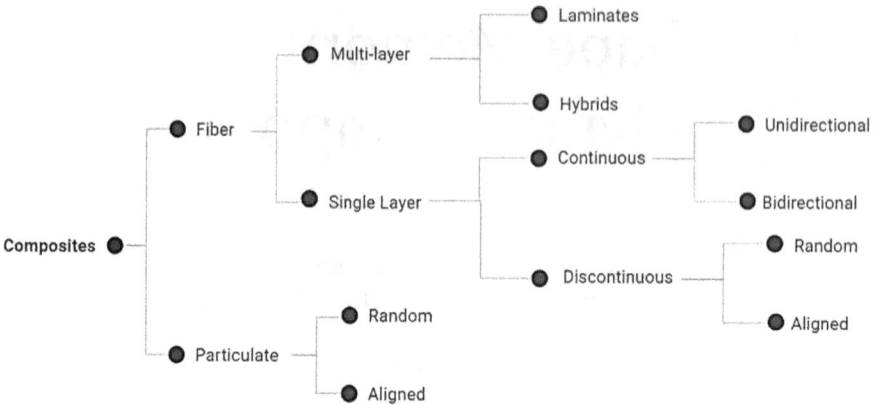

FIGURE 2.1 Classification of composites.

The majority of the composite materials that are addressed in this book are polymer composites, in which the polymer serves as the matrix with different reinforcements. Composites can be divided into fiber composites and particulate composites depending on the size of the support (Figure 2.2). Fibers with a length significantly greater than their cross section are used as reinforcement in fibrous composites or fiber-reinforced polymer composites. The length will be measured in millimeters, centimeters, or meters, and the cross-sectional measurement will be in the micro/nano range. The dimension of the support will be in the range of a few millimeters in particulate composites and micrograms in powdered composites. One of the reinforcement constituents in nanocomposites will be in the nanophase.

Based on the total quantity of the layers, fiber-reinforced polymer composites (FRP) can be divided into single- and multilayer categories. The number of layers can vary depending on the required thickness. Fiber composites can also be

Particulate Composite

Fiber Composite

FIGURE 2.2 Schema of the particulate and fibrous composites (Jones 2018).

divided into continuous fiber composites and short fiber composites based on the length of the fibers. Composites of long continuous fibers produce orthotropic properties, whereas composites of short fibers with random alignment will produce quasi-isotropic properties. Continuous strands can be unidirectional (which uses unidirectional mats), bidirectional, or tridirectional when used as reinforcement. The anisotropy of the compounds will be regulated by the layer direction. When using multilayers to create composites, the layers may be constructed from the same substance or from various elements. Hybrid composites are defined as composites that are constructed from layers of different elements. When it comes to particulate composites, the particles can be distributed randomly or in a coordinated pattern within the matrix system. Compounds with random orientation have quasi-isotropic characteristics, while those with aligned dispersion have orthotropic or anisotropic properties.

2.2 FIBER

The main components of a fiber-reinforced composite substance are fibers. A major characteristic of fiber is its extremely high length-to-diameter ratio. Their effectiveness is typically measured using the strength-to-density and stiffness-to-density ratios. Generally, fibers make up the largest share of the volume in a composite laminate and distribute the bulk of the weight placed on the composite construction. The essential characteristics of a composite laminate, such as density, tensile and compressive strength, fatigue strength, electrical and thermal conductivities, and cost, are affected by the type of fiber, volume percentage, length, and alignment that are chosen.

Requirement of Fiber

The following conditions must be met by fibers before they can be used as support in composites:

- Fibers should have high stiffness and strength;
- Fibers should be 1D (length > cross-sectional dimension);
- Fibers should transmit stress efficiently;
- Fibers should aid composites in having better mechanical features and improved sustainability.

2.2.1 Glass Fibers

Of all the strengthening fibers used in polymeric matrix composites, glass fibers are the most prevalent. The cost of glass fiber is low as compared to Kevlar or carbon fiber, but it still possesses superior strength under compressive and tensile loading, good resistance to chemical corrosion, and insulation capacity. Among industrial fibers, their drawbacks include comparatively low tensile modulus, high density, susceptibility to handling abrasion (which frequently reduces tensile strength), low fatigue resistance, and high hardness (which causes excessive wear on molding dies and cutting tools).

E-glass and S-glass are the two kinds of glass fibers predominantly used in the fiber-reinforced plastics (FRP) business. Apart from these, C-glass is used because it offers higher corrosion resistance to acids for chemical uses. E-glass is used extensively in the FRP production because it is widely accessible and the least expensive reinforcing fiber. S-glass, which was initially created for missile casings and airplane parts, has the greatest tensile strength of all currently used fibers. It is more costly than E-glass due to the compositional variation and greater manufacturing expense. Another variant of S-glass, S-2-glass, is less expensive and is produced to less exacting nonmilitary standards. However, it has tensile strength and stiffness that are comparable to S-glass. Table 2.1 provides a summary of the mechanical characteristics of the E-glass and S-glass fibers.

2.2.2 CARBON FIBER

Carbon fiber tensile modulus values vary from 207 GPa on the low end to 1,035 Gpa on the high end and are readily available. Low-modulus fibers are an alternative to high-modulus fibers because they are lighter, cheaper, stronger, and more resistant to strain failure under tensile loads. Carbon fibers have several advantageous properties, including high fatigue strength, low change in length with unit temperature variation, high tensile strength, and good conductivity to heat. High electrical conductivity, low impact resistance, and low strain-to-failure are a few of its limitations. Their expensive price has so far prevented them from being widely used in commerce. They are primarily used in the aircraft and military sectors, where weight reduction is valued more highly than expense. Properties of the different types of carbon fiber are compiled in Table 2.2.

2.2.3 ARAMID FIBER

Aramid fibers are extremely crystalline aromatic polyamide fibers. Among the present reinforcing fibers have the lowest mass per unit volume and the greatest tensile strength-to-weight ratio. One of the aramid fabrics sold on the market goes by the brand name Kevlar 49. In many maritime and aircraft applications, where light weight, high tensile strength, and resilience to impact damage are essential, aramid fibers are used as reinforcement. They share a negative lengthwise coefficient of thermal expansion, which is similar to that of carbon fiber. This feature helps create

TABLE 2.1
Properties of the E-Glass and S-Glass Fibers

Properties	E-Glass	S-Glass
Specific gravity	2.6	2.5
Modulus (GPa)	72	87
Strength (MPa)	3,450	4,310
Percentage tensile elongation	4.8	5.0
Coefficient of thermal expansion (μm/m/°C)	5.0	5.6

TABLE 2.2
Properties of the Different Types of Carbon Fibers

Properties	Low	Intermediate	Ultrahigh
Specific gravity	1.8	1.9	2.2
Modulus (GPa)	230	370	900
Strength (MPa)	3,450	2,480	3,800
Percentage tensile elongation	1.1	0.5	0.4
Coefficient of thermal expansion (μm/m/°C)	−0.4	−0.5	−0.5

low-thermal-expansion composite panels. Aramid fiber-reinforced composites have the primary disadvantages of weak tensile strengths and challenging manufacturing requirements. Table 2.3 shows the properties of the commonly used aramid fibers.

2.2.4 EXTENDED-CHAIN POLYETHYLENE FIBERS

Long-chain polyethylene strands are made by gel spinning a high-molecular-weight polyethylene and are offered for sale under the brand name "Spectra". Compared to melt spinning, which is used to make traditional polyethylene fibers, gel spinning produces a highly directed fibrous structure with extremely high crystallinity (95%–99%).

The strongest strength-to-weight ratio of any industrial fibers currently on the market is found in Spectra polyethylene fibers. Other notable qualities of Spectra fibers include their low moisture absorption (1% versus 5%–6% for Kevlar 49) and high abrasion resistance, which make them useful as maritime materials in boat hulls and water skiers.

Even at low temperatures, Spectra fibers can provide high-impact protection for composite laminates, and these materials are increasingly being used in ballistic composites like armor, helmets, and other protective gear. However, unless they are combined with stiffer carbon fibers to create hybrid laminates with better impact damage tolerance than all-carbon fiber laminates, their use in high-performance aircraft composites is restricted. Table 2.4 shows the properties of a few Spectra fibers.

TABLE 2.3
Properties of the Different Types of Aramid Fibers

Properties	Kevlar 149	Kevlar 49	Kevlar 129	Kevlar 29
Specific gravity	1.44	1.44	1.44	1.44
Modulus (GPa)	186	124	96	68
Strength (MPa)	3,440	3,700	3,380	2,930
Percentage tensile elongation	2.5	2.8	3.3	3.6
Coefficient of thermal expansion (μm/m/°C)	−2.0	−2.0	−2.0	−2.0

TABLE 2.4

Properties of the Different Types of Spectra Fibers

Properties	Spectra 900	Spectra 1000	Spectra 2000
Specific gravity	0.97	0.97	0.97
Modulus (GPa)	70	105	115
Strength (MPa)	2,600	3,200	3,400
Percentage tensile elongation	3.8	3.0	3.0
Coefficient of thermal expansion $\times 10^6$ (μm/m/°C)	>70	>70	–

2.2.5 NATURAL FIBERS

Pineapple, palm, jute, remi, sisal, coconut fiber (coir), and banana fiber (abaca) are a few examples of natural fibers. All these farming plants are cultivated all over the world for their fabrics, which are frequently used to make ropes, textile backing, bags, and other things. Cellulose microfibrils distributed in an amorphous framework of lignin and hemicellulose make up natural fibers. The cellulose composition of natural fibers ranges from 60 to 80 wt%, and the lignin content is between 5 and 20 wt%, depending on the variety. Additionally, natural fabrics contain up to 20 wt% of moisture. Table 2.5 lists the characteristics of a few of the natural fabrics currently in use.

2.3 MATRIX

The functions of a matrix in composite materials include holding the fibers in place, transferring stresses between the fibers, acting as a barrier against harmful elements like chemicals and moisture, preventing mechanical deterioration of the fiber surface, and providing lateral support to prevent fibers sudden sideways deflection (buckling) under compressive force. In addition to those functions, it also affects the composite's compressive, inter-laminar, and in-plane shear characteristics. The requirements of the matrix are depicted in Figure 2.3. Basically, matrix may be classified into three major categories as: Polymer, Metal, and Ceramic. Further polymer matrix will be sub-classified as thermoplastic (upon heating became soft and became hard whenever cooled): PIE, PVC, nylon, PET, polycarbonate, PEEK and thermosetting type (does not soften upon heating because of the cross-linked nature of polymer molecules): urethane, epoxy, phenolic, and vinyl ester. Metallic matrix uses metal or

TABLE 2.5

Properties of the Different Natural Fibers

Properties	Hemp	Flax	Sisal	Jute
Density (g/cm³)	1.48	1.4	1.33	1.46
Modulus (GPa)	70	60–80	38	10–30
Strength (MPa)	550–900	800–1,500	600–700	400–800
Percentage tensile elongation	1.6	1.2–1.6	2–3	1.8

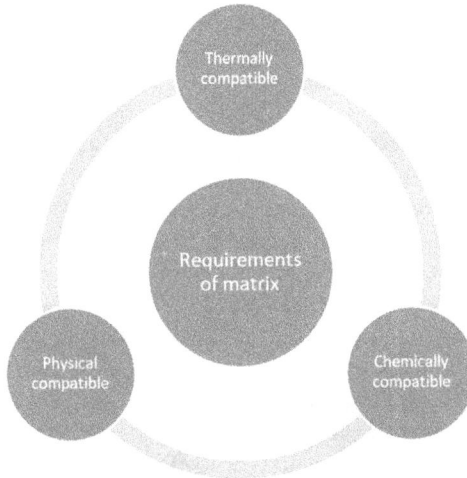

FIGURE 2.3 Basic requirements of the matrix in a composite.

alloy as a matrix. In structural applications lighter metal matrices like magnesium, titanium, aluminum while in high-temperature applications, cobalt, cobalt-nickel alloy, and magnesium matrices are commonly used. In Table 2.6, the mechanical characteristics of a few popular polymer matrixes are summarized.

2.4 SHAPE MEMORY POLYMER COMPOSITES

SMP has exceptional shape recoverability, a straightforward shaping process, low density, and cost-effectiveness, as well as easy temperature management for recovery (Ratna and Karger-Kocsis 2008). Nonetheless, several real-world applications that involve certain functionalities, such as high strength, recovery force, and electrical conductivity, cannot be fulfilled by pure SMPs. In general, the significance of composite materials is attributed to their ability to create synergies and improve the functionality of the primary material by combining multiple components in a harmonious manner. When it comes to SMP, composite materials are developed with two primary objectives: enhancing strength and discovering innovative and efficient stimulus techniques. Through SMPC, the stimulus can be more precise and selective, which can result in the attainment of complex shape memory behaviors. A potential area of study is multifunctional and intelligent SMP materials, which have characteristics such as self-healing (Rodriguez, Luo, and Mather 2011), drag reduction (Chen and Yang 2014), and optical properties (Dong et al. 2014). This section discusses the use of SMP composites, which can serve as a framework, filler, or layer in composites. To enhance their properties, such as reinforcement, self-healing, and other functions, zero-, one-, or two-dimensional carbon materials, polymers, metal and ceramic particles, liquids, or gases can be added. Additionally, SMP can be used as a useful filler in the form of strands or similar structures to create a versatile composite material. The appropriate type and proportion of filler are critical for creating the desired properties. However, adding too much filler can reduce the thermal and mechanical characteristics and lower the transition point. Most fillers also tend to decrease a polymer's shape memory effect, as seen in the fixation rate and recovery rate. Nonetheless, a well-prepared composite framework can accomplish two-way healing and other intriguing and practical tasks (see Figure 2.4).

SMPs are often reinforced to improve their recovery process and carrying capacity, which can be achieved through an increase in both the elastic modulus and general elastic modulus. Reinforcement fillers typically improve both indicators simultaneously, with common options including particle, nanofiber, short-fiber, and long-fiber reinforcements. Carbon materials are a popular choice for reinforced SMPCs due to their strong reinforcement capabilities, versatile physical and chemical properties, and compatibility with polymer-based materials. For example, carbon nanotubes (CNTs), graphene, and carbon fibers have been widely used in these composites. Carbon materials can also form self-assembled structures, such as CNT arrays and carbon nanopaper, along with glass.

The deflection due to tensile loading of SMPs is often limited by the low tensile strain rate in the in-plane direction of the lengthy fiber, which is generally less than 2%. However, considerable geometric nonlinear deformation of the fiber may occur when the material is bent, leading to compressive deformation, known as a post-buckling phenomenon (Dong and Ni 2015). This allows for greater longitudinal compression and deformation. Long-fiber-reinforced soft matter has a different theoretical basis compared to traditional reinforcement methods (Yu et al. 2017). It utilizes buckling and post-buckling to integrate deformation and structure into one material. This strategy has resulted in the production of long-fiber SMPCs, which have been used in the construction of spatially movable structures and actuators, such as spatially deployable hinges, trusses, and antennae (Lu et al. 2010; Zhang et al. 2014).

The shape memory effect of SMPs can be reduced by the addition of an excessive amount of filler or a filler that has an excessively strong reinforcement effect. Although too much or too powerful reinforcement can improve the mechanical characteristics of the polymer, it tends to decrease its shape memory capacity. The reinforcing material typically undergoes elastic deformation after the addition of the reinforcing filler, which results in a lower fixation rate during the cooling process. Furthermore, the reinforcing filler may experience plastic deformation during significant distortion, leading to persistent plastic deformation and a lower healing rate. It is also uncertain how manipulations may affect the temperature at which glass transitions occur. For example, the addition of spherical nanoparticles (NPs) like Ni powder to epoxy-based

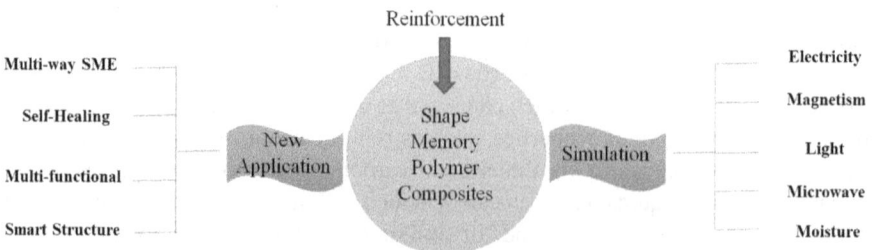

FIGURE 2.4   Effect of reinforcement in SMPCs (Mu et al. 2018).

TABLE 2.6

Properties of Some Polymer Matrix

Properties	Epoxy	Poly Ester	Vinyl Ester	Phenolic	Polyphenylene Sulfide (PPS)	Polyamide-Imide (PAI)
Tensile strength (MPa)	55–130	34–105	73–81	50–60	82.7	102
Tensile modulus (GPa)	2.7–4.1	2.1–3.5	3-3.5	4–7	3.3	3.03
Flexural strength (MPa)	110–150	70–110	130–140	80–135	152	212
Tensile modulus (GPa)	3–4	2–4	3	2–4	3.45	4.55

SMPs can lower the glass transition temperature (Leng et al. 2008), whereas the doping of styryl SMPs with CNF/MWCNT has the opposite effect (Lei et al. 2016).

Meng et al. (2017) conducted an experiment to assess the shape memory properties of SMP by incorporating C60, CNTs, and graphene into them. Material possess its original shape at room temperature. Whenever it is heated, one can change the shape of the material or mold the material into the required shape and cool it again, and this is called a programmed shape. But the material regains its original shape and size after reheating above T_g. But it is observed that full recovery was not achieved with the modification of polymers by the addition of graphene, and that will increase stresses in the material, which may be termed as recovery stress. But the case with C60 and CNT is different. Here, with the addition of C60 and nanotubes, the recovery rate percentage is increased significantly but has little effect on the shape recovery stress. It is crucial to consider the application requirements when balancing the appropriate shape and proportion of the filler. Some fillers can serve dual purposes, providing both shape-recall impact and reinforcement. These fillers typically rely on supramolecular structures. For instance, cellulose whiskers and graphene oxide are reinforcing materials that interact with the polymer network through hydrogen bonds, resulting in a mutable supramolecular effect to establish the desired shape (Qi et al. 2014). Furthermore, incorporating self-organized nanocrystals such as carbon black into the polymer can improve the material's fixation and healing rates (Qi et al. 2017).

2.4.1 SHAPE MEMORY FIBER-REINFORCED POLYMER COMPOSITES

When it comes to increasing the mechanical strength of SMPs, chopped and continuous microfibers and fabrics, and particularly continuous microfibers and fabrics, are preferable to micro- and nano-sized particles. It is not possible to attain excellent SME in the fiber orientation of the composite due to the high elastic modulus of fibers or textiles. The shear modulus of elasticity of a composite material can typically be estimated by bending the material in a direction that is transverse to the orientation of the fibers (Tiwari and Shaikh 2019, 2021a). SMPs that have been strengthened by microfiber or fabric are primarily used for two applications: devices that allow spaceships to launch and devices that regulate vibration. Elastic memory composites are another name for fiber/fabric-strengthened SMPs that are used for applications involving self-deployable space constructions (Sokolowski and Tan 2007). Elastic memory composites have the ability to be compressed on earth and then transported to space in their compressed state. In space, they can self-deploy. The management of vibration is another application for fiber and fabric-incorporated SMPs. SMP composites have the advantages of high energy absorption effectiveness, low density, and high shape deformability, making them ideal for use as vibration control materials.

Researchers conducted a number of experiments to investigate the effects of reinforcing thermoplastic shape memory polyurethane with different types of fibers and fabrics. The results indicate that chopped fiberglass, unidirectional Kevlar fiber, and woven fiberglass can significantly increase the strength and stiffness of SMPs (Ohki et al. 2004; Liang, Rogers, and Malafeew 1997). Similarly, experiments conducted on epoxy laminates strengthened with carbon fiber, glass fiber, and Kevlar fiber have shown that the use of microfibers or textiles is an effective way to increase the stiffness of epoxy SMP resins (Gall et al. 2000).

The microbuckling effect was found to be the primary cause of the bending strain in thermoset SMP resin based on styrene and strengthened by fabric with a unidirectional pattern. Moreover, work on carbon fiber fabrics reinforced shape memory polyurethane laminates has revealed that the arrangement and location of shape memory polyurethane films and carbon fiber fabrics in laminates have a significant impact on the shape memory characteristics (Zhang and Ni 2007). In addition, research has been carried out on the mechanisms of failure and the accuracy of implementation of elastic memory composites by various researchers (Campbell and Maji 2006). Overall, the research suggests that the use of fibers and fabrics can substantially improve the mechanical properties of SMPs and their shape memory characteristics.

Lan et al. utilized a thermoset styrene-based SMP reinforced with carbon-based plain-weave fabrics to create a self-deployable hinge (Lan et al. 2009). The hinge consisted of two cylindrical SMP composite casings with a curved shape and was triggered when an appropriate voltage was applied to the integrated resistor. It got warmer in each laminate. SMP composites have been shown to be effective in controlling vibration by absorbing vibration energy through structure displacement at temperatures close to their glass transition temperature (T_g). In another study, Yang et al. developed a structural composite whose sheet was made of shape memory polyurethane and an epoxy beam, which served as a vibration control material (Heung Yang et al. 2004). Compared to an epoxy beam alone, the resistance against sudden impact of the composite material was up to four times greater.

These studies highlight the potential of SMP composites in developing smart materials with improved mechanical properties, such as self-deployable hinges and impact-resistant vibration control substances. By utilizing SMP composites, it is possible to create innovative materials that can respond to changes in temperature, electrical stimulation, or other stimuli, making them beneficial in a variety of applications, including aerospace, automotive, and biomedical industries.

2.4.2 SHAPE MEMORY POLYMER NANOCOMPOSITES

Over the past few years, researchers have developed SMPNCs as a way to address the restrictions of traditional SMPs. Compared to classical composite fillers, nanofillers have proven to be more effective due to their larger surface area and stronger interaction with the polymer. To produce SMPNCs, one or more nanofillers, such as nanotubes/fibers/spheres/rods are incorporated into the matrix. This allows for the creation of specific shape memory effects, including the triple shape memory effect, and the development of new shape memory functions using non-shape memory materials (Tiwari and Shaikh 2021b, 2021c). Additionally, to activate thermally sensitive SMPs, nonthermal changes like electricity, magnetic fields, light, humidity, water, or solvent may be introduced. The incorporation of nanofillers has also been shown to enhance various SMP properties, such as fixity ratio, strain recovery ratio, shape recovery stress, and recovery speed, as well as broaden the actuation range and tune the switching temperature (Tiwari and Shaikh 2022a, 2022b).

To activate SMPs, an external heat source is required, and because of the gentle heat transmission and response time of conventional SMPs, their actuation became difficult. In that case, there are certain nanofillers, and with the addition of these, the polymer conducting capacity will be raised. Figure 2.5 shows some of the conductive nanofillers that will raise the conductivity of polymers with their addition in a very small percentage. These fillers generate heat based on Joule's law, which facilitates heat transfer and triggers the shape memory effect in the SMPs.

2.4.2.1 Carbon Nanotubes (CNTs) Based SMPNCs

Previous literature studies show that because of its exceptionally high modulus, strength, stiffness, aspect ratio, and conducting nature, nanotube addition in polymers became ideal (Coleman, Khan, and Gun'ko 2006). While creating conductive polymer compounds, high electrical and thermal transfer capability of CNTs is very

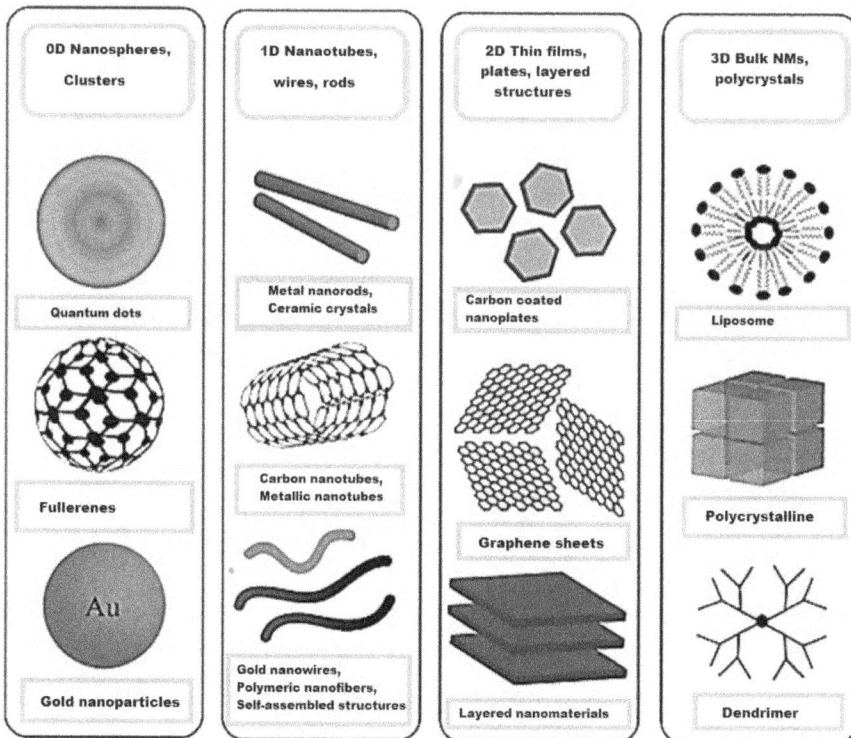

FIGURE 2.5 Classification of the nanoparticles (Alkaç et al. 2021).

helpful (Han and Fina 2011). A remarkable nucleation impact on the crystallization of semi-crystalline polymers is also demonstrated by CNTs (Xu et al. 2014). This is due to the fact that CNTs have a substantial surface area and a chain-like structure that resembles the structure of polymer molecules in some ways.

CNTs are known for their exceptional mechanical properties, which arise from carbon–carbon sp2 bonding. CNTs have high stiffness and axial strength, with an average Young's modulus of isolated single-walled nanotubes (SWNTs) estimated to be 1.25 ± 0.40 terapascals (TPa) (Krishnan et al. 1998). This value is significantly higher than that of carbon fibers, which is approximately 680 gigapascals (GPa) (Jacobsen et al. 1995). The estimated Young's modulus of multiwalled nanotubes is 1.26 TPa (Wong, Sheehan, and Lieber 1997).

The nature of CNT is quite versatile; it can get stretch, twist about an axis, get rolled, or form a flat plate before undergoing breakage or fracture. It is observed that its flexural and fracture strengths are 14.2 ± 8.0 and 11–63 GPa, respectively, which will obviously depend on shell diameter (Yu, Lourie, et al. 2000). The theoretical average Young's and shear moduli of SWNT ropes are about 1.0 and 0.5 TPa, respectively (Lu 1997).

The effective modulus of the single-walled CNT is about 1,000 GPa, and that of diamond is 1,200 GPa; also, their strengths are 10–100 times higher than steel (Vita et al. 1999). The nanotubes have a breaking strain of up to 5.3%. The stress–strain curves of SWNT ropes show that the load is primarily carried by SWNTs on the border of the ropes, resulting in breaking strengths ranging from 13 to 52 GPa (Yu, Files, et al. 2000). However, this is much less than that described for a single multiwalled nanotubes, which can buckle elastically at large deflection angles of approximately 10° for lengths of 1 μm (Wang et al. 2001).

The mechanical characteristics of CNTs, such as Young's modulus, Poisson's ratio, and bulk modulus, are dependent on the CNT tube radius and the technique of measurement. A summary of the mechanical properties of CNTs is presented in Table 2.7.

2.4.2.2 Graphene Based SMPNCs

A 2D film of carbon molecules with sp^2 bonding makes up graphene. It has high heat conductivity (5,300 WmK^{-1}), E~1,086 GPa, high carrier mobility (200,000 cm^2/Vs^1), low optical absorption (~2.3%) in the visual range, and robust near-infrared light absorption (Stankovich et al. 2006). Graphene has a tendency to aggregate and take on the graphite structure as a consequence of the van der Waals force of attraction, which clues to the inefficient dispersion of graphene

TABLE 2.7

Mechanical Properties of CNTs

Material	Tensile Strength (GPa)	Young's Modulus (TPa)	Measurement Technique
SWNT rope	45 ± 7	0.32–1.47	Micro-Raman spectroscopy
Buckypaper of SWNT	0.030	0.008 (at 0.2% strain)	Tensile-loading experiment (elongation to break 0.5%)
Multiwalled nanotubes	11–63	0.27–0.95	Atomic force microscopy

into a polymer matrix. The agglomeration issue can be resolved by functionalizing grapheme oxide, oxidizing graphene, and using reduced graphene oxide (Choi et al. 2010). Large oxygen-containing groups, such as carboxyl, hydroxyl, and epoxide groups, are present on the margins and basal surfaces of GO. For the purpose of functionalizing GO, two distinct approaches are used: the first involves functionalizing the basal plane using the reactivity of hydroxyl and epoxy groups, while the second involves functionalizing the margins using the reactivity of carboxylic groups (Choi et al. 2010). Some SMPNCs built on graphene-based NPs are listed in Table 2.8.

Mechanical behavior of a graphene monolayer hung over a hole on a silicon base is shown as a stress–strain curve in Figure 2.6, measured by an atomic force microscope using a nanoindentation. It is clear that as strain increases and the crack happens at a strain of over 25%, the stress–strain graph becomes nonlinear. The defect-free graphene has a breaking strength of 130 ± 10 GPa and a Young's

TABLE 2.8

Examples of Graphene-Based SMPNC (Bhanushali et al. 2022)

S. No.	Matrix	Nanoparticle	Graphene Content	Stimulus Field
1.	Segmented polyurethane	Reduced GO	1 phr	Thermal
2.	Polyurethane	GO	–	Thermal
3.	Castor oil-based polyurethane	Graphene nanoplatelets	–	Thermal/Electrical
4.	Shape memory polyurethane	Graphene nanosheets	2 wt%	Thermal/Electrical
5.	Polyurethane	GQD	–	Thermal/Electrical
6.	Shape memory epoxy polymer (SMEP)	GO	2 wt%	Thermal
7.	Epoxy/cyanate ester	Graphene nanoplatelets	2.4 wt%	Thermal/Electrical
8.	Poly lactic acid	Graphene nanoplatelets	4 wt%	Thermal
9.	Poly(vinyl alcohol)	GO	1 wt%	Moisture

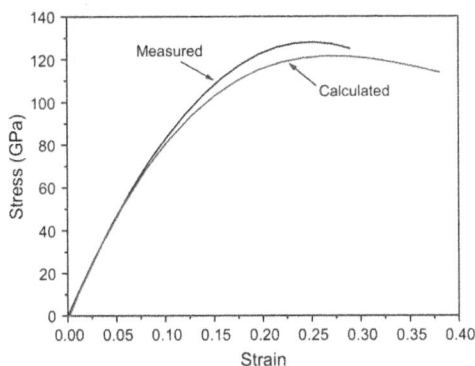

FIGURE 2.6 Elastic behavior of the graphene monolayer through the stress–strain relationship (Young et al. 2012).

modulus of $1,000 \pm 100$ GPa. These numbers are quite similar to those anticipated theoretically by density functional theory (Liu, Ming, and Li 2007). Table 2.9 presents a compendium of the characteristics shared by many of the most common allotropes of carbon NPs.

2.4.2.3 SMPNCs with Cellulose Based Nanoparticles

Cellulose is a substance that is harvested from plants, bacterial cell walls, and other biological things. Cellulose nanocrystals, a comparatively novel class of cellulose-based NPs, are composed of strands of cellulose packed tightly with numerous OH groups on their surfaces. As a result, cellulose nanocrystals and matrix materials engage strongly through hydrogen bonds. With diameters of about 8–10 nanometer (nm) and lengths of 100–200 nm, cellulose nanocrystals have a rod-like form with high aspect ratios, making them a more sustainable, biodegradable, and reusable option to carbon-based fillers (Chen et al. 2018). To achieve a consistent distribution of polar nanocellulose within a nonpolar matrix, the solvent casting technique is typically used (Huq et al. 2012).

2.4.2.4 Titanium Nitride (TiN)Based SMPNCs

NPs may significantly affect light transmission, despite their size being much smaller than the wavelength of a light wave. The most well-known characteristic of NPs is the

TABLE 2.9
Properties of Some Nanocarbon Allotropes (Wu et al. 2012)

Carbon Allotropes	Graphite	Diamond	Fullerene	Carbon Nanotubes	Graphene
Hybridized form	sp^2	sp^3	sp^2	sp^2	sp^2
Crystal structure	Hexagonal	Cubic	FCC 1	FCC	Hexagonal
Dimensions	Three	Three	Zero	One	Two
Experimental specific surface area (m^2/g)	~10–20	20–160	80–90	~1,300	~1,500
Density (g/cm)	2.09–2.23	3.5–3.53	1.72	>1	>1
Optical properties	Uniaxial	Isotropic	Nonlinear optical response	Structure dependant	97.7% of optical transmittance
Thermal conductivity (W/mK)	1,500–2,000 5–10	900–2,320	0.4	3,500	4,840–5,300
Hardness	High	Ultrahigh	High elastic	High-flexible elastic	Highest (single layer)
Tenacity	Flexible nonelastic	–	Elastic	Flexible elastic	Flexible elastic
Electronic properties	Electrical conductor	Insulator	Insulator	Metallic and semiconducting	Semimetal, zero-gap semiconductor
Electrical conductivity	Anisotropic, 2.3×10^4, 6b	–	10^{10}	Structure dependant	2,000

confined surface plasmon tone that can be generated in metallic NPs, traditionally using gold (Au) and silver (Ag). However, recent research has revealed that plasmonic resonances can also be generated in evolution metal nitrides such as titanium nitride (TiN) and zirconium nitride (ZN), as well as highly altered translucent conductive oxides (West et al. 2010; Naik, Shalaev, and Boltasseva 2013).

Plasmonic resonances can be excited by a variety of elements and materials, not just metal NPs. TiN NPs have a wider plasmonic frequency range than Au or Ag NPs due to their greater optical absorbance loss. In addition, TiN NPs have a higher absorption efficiency than other commonly used light-absorbing NPs, such as carbon NPs, which are typically used in the same amounts (Neumann et al. 2013; Han et al. 2011). The resonance point of TiN NPs is approximately 700 nm, allowing them to cover near-infrared light through plasmon-enhanced absorbance (Guler et al. 2012).

2.4.2.5   Titanium Dioxide (TiO$_2$)Based SMPNCs

Rutile, brookite, and anatase are three different polymorphs of TiO$_2$ that have unique material properties like electric, magnetized catalytic in nature, and electrical properties (Mo and Ching 1995). Rutile TiO$_2$ is used as a bleaching agent because it has a broad band-gap, a high refractive index, hiding power, excellent chemical stability, and UV light screening effects (Mehranpour et al. 2010). TiO$_2$ can function effectively as an opacifier when used in power form (Ingham et al. 2009).

2.4.2.6   Nanoclay Based SMPNCs

According to their molecular makeup and shape, the stratified mineral silicate NPs known as nanoclays are classified as montmorillonite, bentonite, kaolinite, hectorite, and halloysite. Due to its cheap price, high aspect ratio, and high mechanical and thermal characteristics, nanoclay has received a great deal of attention recently (Patel, Biswas, and Maiti 2016). However, due to its excessive hydrophilicity, nanoclay is difficult to disperse into an organic polymeric phase (Biswas, Aswal, and Maiti 2019).

2.4.3   Shape Memory Hybrid Composites

Hybrid material systems are created by combining two or more different materials or types of reinforcement at the molecular or nanometer level. These systems are known to be stiffer and stronger than two-phase/regular reinforced polymer composites. In a mixed-material system, one type of fiber usually has a lower modulus and price, while the other type has a high modulus and cost (Safri et al. 2018). By combining multiscale components, hybrid material systems can exhibit new properties that possess advanced and improved characteristics compared to individual materials or parts. Hybrid material systems have a lot of promise for making complex-shaped structures that are light, strong, and rigid. These parts are often used in the automobile and aircraft industries (Okafor et al. 2023, 2022). It's crucial to have a comprehensive understanding of how individual components interact within a hybrid material system for efficient future design.

The specific moduli of graphite, Kevlar, and glass fibers are 227.5, 124.11, and 85 GPa, respectively. Cost-wise conventionally, Kevlar fiber is more expensive than other fibers. But if one wants to utilize opportunity provided by each individual, then they can use layers of different fiber to form a layered composite. That method of preparation of hybrid composite provides great strength, stiffness, and effective load-absorbing capacity, along with reducing the overall cost of structural composite (Dong et al. 2016). GLARE is one example of such a hybrid composite. It consists of layers of aluminum sheets, which are further sandwiched between a glass fiber-reinforced composite face sheet and a bottom sheet. Such composites find their application in the Airbus A380 fuselage. The methods of preparation of hybrid composites were not limited to this. Instead of multiple fibers, one can also use matrix–hybrid composite.

Hybrid composites can be classified as either matrix–hybrid composites or fiber–hybrid composites. Matrix–hybrid composites have thermoplastic fibers that dissolve into a thermosetting matrix during curing, whereas fiber–hybrid composites have thermoplastic fibers that retain their fiber shape after curing. Hybrid composites' ability to absorb energy is greatly aided by the plastic deformation of thermoplastic fibers. Hybrid composites with thermoplastic fibers that melt into the thermosetting matrix absorb almost as much energy as plain glass fiber composites, suggesting that fiber properties are more important than matrix properties in determining total energy absorbed. The overall amount of energy absorbed by hybrid composites varies somewhat depending on the matrix. Hybrid composites' performance is a compromise between the advantages and disadvantages of the component fibers. The respective volume fractions of the two fibers, the loading configuration, and the fiber layout all play a role in determining the nature of the hybrid effect. Hybridization may also be accomplished by the use of a combination of natural and synthetic fibers (Rajkumar, Srinivasan, and Suvitha 2015).

The purpose of combining two distinct fibers into a single composite is to improve upon the positive qualities of each while compensating for the negative ones. It is possible to considerably lower prices while preserving virtually the same flexural qualities by, say, substituting pricey carbon fibers with cheaper glass fibers in the center of a laminate. The more brittle fibers in a hybrid composite tend to break under stress in the fiber direction before the more ductile fibers do. Both Czél and Wisnom (2013) and Huang et al. (2006) note that this fracture behavior may be used for health monitoring or as a warning indicator before ultimate failure. The LE fiber often fails before the HE fiber does; hence, the two are sometimes referred to as "low-elongation" and "high-elongation" fibers, respectively. The failure strain of the HE fiber is always going to be higher than that of the LE fiber, even if the difference isn't huge. This is also why it's not always clear what's meant when "brittle" and "ductile" are used instead of LE and HE fibers.

There are many ways in which the LE and HE fibers can be combined to create a hybrid composite. The most commonly used configurations are shown in Figure 2.7. The first method, referred to as the preparation of a number of stacks of fibers over each other, as shown in Figure 2.7a, involves stacking layers of the two fiber types on top of each other. Producing a hybrid composite material in this manner results in the fewest complications and the lowest overall cost. The intralayer configuration looks very similar to MAT-type structures, grids, or weaving of different fibers as shown in Figure 2.7b. Additionally, it is possible to mix or combine the two different types of fibers at the fiber level, which will result in an intrayarn hybrid (Figure 2.7c).

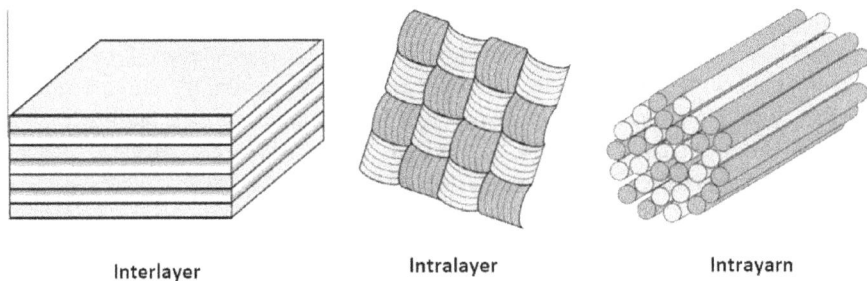

Interlayer   Intralayer   Intrayarn

FIGURE 2.7   Different hybrid configurations in fiber-reinforced composites (Swolfs, Gorbatikh, and Verpoest 2014).

Combining two of these three different arrangements may result in the creation of more complicated configurations. For instance, a homogenous yarn and an intrayarn hybrid may be woven together to create a single fabric.

When manufacturing hybrid composites, the dispersion of the various types of fibers is an essential factor to take into consideration. According to Manders and Bader (1981), this describes the degree to which the two types of fiber are combined successfully and is quantified as the reciprocal of the length of the pattern with the shortest repetition. The various degrees of dispersion are shown graphically in Figure 2.7. A low degree of dispersion is present in interlayer hybrids, which organize the two different kinds of fiber into separate layers. As shown in intralayer hybrids, this may be enhanced by either increasing the number of layers or lowering the layer thickness (Czél and Wisnom 2013). Hybridization on the level of the fiber bundle, as seen in intrayarn, is still another method that may be used to increase the dispersion. Therefore, the best dispersion is achieved when the two different types of fiber are distributed in an entirely random manner.

The concept of hybrid effect may be seen in Swolfs, Gorbatikh, and Verpoest (2014); however, in order to fully understand this definition, it is important to precisely determine the failure strain of the composite made from carbon fiber. Accumulation of stresses at the grips can affect the baseline failure strain, but this effect is smaller in hybrid composites (Kretsis 1987). Overestimations of the hybrid effect can occur as a result. Additionally, it is important to note that calculating the hybrid effect based on the ultimate failure strain of the hybrid composite is not accurate. While improvements in ultimate failure strain may be worth reporting, the term "hybrid effect" should not be used to describe them.

Based on displacement-controlled testing where isostrain is assumed for both low-energy (LE) and high-energy (HE) fibers, some writers have proposed that the hybrid effect for tensile strength follows a bilinear rule of mixing. As a simplification, we ignore the matrix's effect. When the LE fibers in this model finally give out, the HE fibers are left to carry the whole weight by themselves. After the failure of LE fibers, two possibilities emerge, depending on the proportion of HE fibers. The stress contribution of the HE fibers at their failure strain may become dominant at large percentages of HE fibers, exceeding the stress at their failure strain of the LE fibers. In contrast, when the proportion of HE fibers is small, the strength is governed by the stress in the hybrid when the LE fibres break because the HE fibers continue to transmit load, but their stress at failure does not exceed the stress at the failure strain of the LE fibers. When the tensile peaks are all the same height, the bilinear rule of mixtures predicts a minimum. However, a positive deviation from the bilinear rule of mixtures is indicated by experimental data from Yentl Swolfs et al. (2019) for carbon/glass hybrid composites and Peijs et al. (1990).

2.5   FIBER CONTENT, DENSITY, AND VOID CONTENT

Theoretical calculations of a fiber-reinforced composite's strength, modulus, and other properties rely on the fiber volume fraction present in the material. However, in practical experiments, it is often simpler to determine the fiber weight fraction (w_f) of the composite. From this weight fraction, the fiber volume fraction (v_f) and the density of the composite (ρ_c) can be calculated by Eqs. 2.1 and 2.2. The density of composites (ρ_c) can also be calculated through $\rho_c = \rho_f v_f + \rho_m v_m$ in terms of volume fraction.

$$v_f = \frac{w_f/\rho_f}{(w_f/\rho_f)+(w_m/\rho_m)} \tag{2.1}$$

$$\rho_c = \frac{1}{(w_f/\rho_f)+(w_m/\rho_m)} \tag{2.2}$$

where
  w_f = weight of fiber in the composite
  w_m = weight of matrix in the composite
  ρ_f = density of fiber
  ρ_m = density of matrix.

Sometimes it may be possible that fibers will smoothly remove from composites and no restriction occurs from the polymer side. This may have occurred due to improper binding of the fiber matrix. To have good bonding between fiber and matrix, the surface of the fiber will be coated with a chemical layer, and this process is nothing but interphase preparation.

The weight fraction of fibers in a composite material can be determined through experimental methods such as the ignition loss method (ASTM D2854) or the matrix digestion method (ASTM D3171). The ignition loss method is used for composites containing fibers that do not lose weight at high temperatures, such as glass fibers. In this method, a small sample of the composite is heated in a muffle furnace at temperatures between 500°C and 600°C to burn off the cured resin, and the weight of the remaining fibers is measured.

In the matrix digestion method, the matrix material, whether polymeric or metallic, is dissolved away using a suitable liquid medium, such as concentrated nitric acid. The weight of the remaining fibers is then measured, and the weight fraction of the fibers is determined by comparing the weight of the test sample before and after the matrix is removed.

For composites containing unidirectional conductive fibers such as carbon in a nonconductive matrix, the fiber volume fraction can be determined directly by comparing the electrical resistivity of the composite with that of the fibers using ASTM D3355.

During the manufacturing of composite materials, air or other volatiles may become trapped in the material, either during the incorporation of fibers into the matrix or during the production of laminates. These trapped air or volatiles can create microvoids in the composite material, which can significantly impact its mechanical properties.

A high void content, typically over 2% by volume, can lead to lower fatigue resistance, increased susceptibility to water diffusion, and increased variation in mechanical properties. The void content in a composite laminate can be estimated by comparing its theoretical density (calculated based on the volume fractions of the constituents) with its actual density as:

$$v_v = \frac{\rho_c - \rho}{\rho_c} \tag{2.3}$$

where
  v_v = volume fraction of voids
  ρ_c = theoretical density of composites
  ρ = experimental density.

If the actual density is lower than the theoretical density, then the composite material contains voids. The void content can be calculated as the difference between the theoretical density and the actual density, divided by the theoretical density. Several methods can be used to measure the actual density, such as immersion in fluids of known density or weighing a known volume of the material. It's important to minimize the void content in composite materials to ensure optimal mechanical performance.

SWNT, single-walled nanotube.
GO, graphene oxide.

REFERENCES

Alkaç, İsmail Mert, Burcu Çerçi, Cisil Timuralp, and Fatih Şen. 2021. "2- Nanomaterials and Their Classification." In *Nanomaterials for Direct Alcohol Fuel Cells*, edited by Fatih Şen, 17–33. Micro and Nano Technologies. Elsevier. https://doi.org/10.1016/B978-0-12-821713-9.00011-1.

Bhanushali, Haresh, Shweta Amrutkar, Siddhesh Mestry, and S. T. Mhaske. 2022. "Shape Memory Polymer Nanocomposite: A Review on Structure–Property Relationship." *Polymer Bulletin* 79 (6): 3437–93. https://doi.org/10.1007/s00289-021-03686-x.

Biswas, Arpan, Vinod K. Aswal, and Pralay Maiti. 2019. "Tunable Shape Memory Behavior of Polymer with Surface Modification of Nanoparticles." *Journal of Colloid and Interface Science* 556 (November): 147–58. https://doi.org/10.1016/j.jcis.2019.08.053.

Campbell, Douglas, and Arup Maji. 2006. "Failure Mechanisms and Deployment Accuracy of Elastic-Memory Composites." *Journal of Aerospace Engineering* 19 (3): 184–93. https://doi.org/10.1061/(ASCE)0893-1321(2006)19:3(184).

Chen, Chi-Mon, and Shu Yang. 2014. "Directed Water Shedding on High-Aspect-Ratio Shape Memory Polymer Micropillar Arrays." *Advanced Materials* 26 (8): 1283–8. https://doi.org/10.1002/adma.201304030.

Chen, Wenshuai, Haipeng Yu, Sang-Young Lee, Tong Wei, Jian Li, and Zhuangjun Fan. 2018. "Nanocellulose: A Promising Nanomaterial for Advanced Electrochemical Energy Storage." *Chemical Society Reviews* 47 (8): 2837–72. https://doi.org/10.1039/C7CS00790F.

Choi, Eun-Young, Tae Hee Han, Jihyun Hong, Ji Eun Kim, Sun Hwa Lee, Hyun Wook Kim, and Sang Ouk Kim. 2010. "Noncovalent Functionalization of Graphene with End-Functional Polymers." *Journal of Materials Chemistry* 20 (10): 1907–12. https://doi.org/10.1039/B919074K.

Coleman, Jonathan N., Umar Khan, and Yurii K. Gun'ko. 2006. "Mechanical Reinforcement of Polymers Using Carbon Nanotubes." *Advanced Materials* 18 (6): 689–706. https://doi.org/10.1002/adma.200501851.

Czél, Gergely, and M.R. Wisnom. 2013. Demonstration of pseudo-ductility in high performance glass/epoxy composites by hybridisation with thin-ply carbon prepreg. *Composites Part A: Applied Science and Manufacturing* 52: 23–30. https://doi.org/10.1016/j.compositesa.2013.04.006

Dong, Junliang, Alexandre Locquet, Nico F. Declercq, and D. S. Citrin. 2016. "Polarization-Resolved Terahertz Imaging of Intra- and Inter-Laminar Damages in Hybrid Fiber-Reinforced Composite Laminate Subject to Low-Velocity Impact." *Composites Part B: Engineering* 92 (May): 167–74. https://doi.org/10.1016/j.compositesb.2016.02.016.

Dong, Yubing, and Qing-Qing Ni. 2015. "Effect of Vapor-Grown Carbon Nanofibers and in Situ Hydrolyzed Silica on the Mechanical and Shape Memory Properties of Water-Borne Epoxy Composites." *Polymer Composites* 36 (9): 1712–20. https://doi.org/10.1002/pc.23082.

Dong, Yubing, Qing-Qing Ni, Lili Li, and Yaqin Fu. 2014. "Novel Vapor-Grown Carbon Nanofiber/Epoxy Shape Memory Nanocomposites Prepared via Latex Technology." *Materials Letters* 132 (October): 206–9. https://doi.org/10.1016/j.matlet.2014.06.084.

Fejős, M., G. Romhány, and J. Karger-Kocsis. 2012. "Shape Memory Characteristics of Woven Glass Fibre Fabric Reinforced Epoxy Composite in Flexure." *Journal of Reinforced Plastics and Composites* 31 (22): 1532–7. https://doi.org/10.1177/0731684412461541.

Gall, Ken, Martin Mikulas, Naseem A. Munshi, Fred Beavers, and Michael Tupper. 2000. "Carbon Fiber Reinforced Shape Memory Polymer Composites." *Journal of Intelligent Material Systems and Structures* 11 (11): 877–86. https://doi.org/10.1106/EJGR-EWNM-6CLX-3X2M.

Guler, U., G. V. Naik, A. Boltasseva, V. M. Shalaev, and A. V. Kildishev. 2012. "Performance Analysis of Nitride Alternative Plasmonic Materials for Localized Surface Plasmon Applications." *Applied Physics B* 107 (2): 285–91. https://doi.org/10.1007/s00340-012-4955-3.

Han, Dongxiao, Zhaoguo Meng, Daxiong Wu, Canying Zhang, and Haitao Zhu. 2011. "Thermal Properties of Carbon Black Aqueous Nanofluids for Solar Absorption." *Nanoscale Research Letters* 6 (1): 457. https://doi.org/10.1186/1556-276X-6-457.

Han, Zhidong, and Alberto Fina. 2011. "Thermal Conductivity of Carbon Nanotubes and Their Polymer Nanocomposites: A Review." *Progress in Polymer Science, Special Issue on Composites* 36 (7): 914–44. https://doi.org/10.1016/j.progpolymsci.2010.11.004.

Heung, Yang, Jae Byoung Chul Chun, Yong-Chan Chung, Jae Whan Cho, and Bong Gyoo Cho. 2004. "Vibration Control Ability of Multilayered Composite Material Made of Epoxy Beam and Polyurethane Copolymer with Shape Memory Effect." *Journal of Applied Polymer Science* 94 (1): 302–7. https://doi.org/10.1002/app.20902.

Huang, Chunyue, Zhaohua Wu, and Dejian Zho. 2006. "Simulation and Analysis of Dynamic Characteristics of Micro-Acceleration Switch with Finite Element Method." In *2006 2nd IEEE/ASME International Conference on Mechatronics and Embedded Systems and Applications*, Beijing, China, 1–5. doi: 10.1109/MESA.2006.296952.

Huq, Tanzina, Stephane Salmieri, Avik Khan, Ruhul A. Khan, Canh Le Tien, Bernard Riedl, Carole Fraschini, et al. 2012. "Nanocrystalline Cellulose (NCC) Reinforced Alginate Based Biodegradable Nanocomposite Film." *Carbohydrate Polymers* 90 (4): 1757–63. https://doi.org/10.1016/j.carbpol.2012.07.065.

Ingham, Bridget, Scott Dickie, Hiroshi Nanjo, and Michael F. Toney. 2009. "In Situ USAXS Measurements of Titania Colloidal Paint Films during the Drying Process." *Journal of Colloid and Interface Science* 336 (2): 612–5. https://doi.org/10.1016/j.jcis.2009.04.035.

Jacobsen, Ronald L., **Terry** M. Tritt, Jason R. Guth, Adam C. Ehrlich, and Delmar J. Gillespie. 1995. "Mechanical Properties of Vapor-Grown Carbon Fiber." *Carbon* 33 (9): 1217–21. https://doi.org/10.1016/0008-6223(95)00057-K.

Jones, Robert M. 2018. *Mechanics of Composite Materials*. 2nd ed. CRC Press: Boca Raton. https://doi.org/10.1201/9781498711067.

Kretsis, George. 1987. "A Review of the Tensile, Compressive, Flexural and Shear Properties of Hybrid Fibre-Reinforced Plastics." *Composites* 18 (1): 13–23. https://doi.org/10.1016/0010-4361(87)90003-6.

Krishnan, A., E. Erik Dujardin, T. W. Ebbesen, Peter N. Yianilos, and Michael M. J. Treacy. 1998. "Young's Modulus of Single-Walled Nanotubes." *Physical Review B* 58 (20): 14013–9. https://doi.org/10.1103/PhysRevB.58.14013.

Lan, Xin, Yanju Liu, Haibao Lv, Xiaohua Wang, Jinsong Leng, and Shanyi Du. 2009. "Fiber Reinforced Shape-Memory Polymer Composite and Its Application in a Deployable Hinge." *Smart Materials and Structures* 18 (2): 024002. https://doi.org/10.1088/0964-1726/18/2/024002.

Lei, Ming, Ben Xu, Yutao Pei, Haibao Lu, and Yong Qing Fu. 2016. "Micro-Mechanics of Nanostructured Carbon/Shape Memory Polymer Hybrid Thin Film." *Soft Matter* 12 (1): 106–14. https://doi.org/10.1039/C5SM01269D.

Leng, Jinsong S., Xinzheng Lan, Y. Yayuan J. Liu, Samuel Y. Du, Wan-Ting Huang, Nian Liu, S. J. Phee, and Quan Yuan. 2008. "Electrical Conductivity of Thermoresponsive Shape-Memory Polymer with Embedded Micron Sized Ni Powder Chains." *Applied Physics Letters* 92 (1): 014104. https://doi.org/10.1063/1.2829388.

Liang, C., C. A. Rogers, and E. Malafeew. 1997. "Investigation of Shape Memory Polymers and Their Hybrid Composites." *Journal of Intelligent Material Systems and Structures* 8 (4): 380–6. https://doi.org/10.1177/1045389X9700800411.

Liu, Fang, Pingbing Ming, and Ju Li. 2007. "Ab Initio Calculation of Ideal Strength and Phonon Instability of Graphene under Tension." *Physical Review B* 76 (6): 064120. https://doi.org/10.1103/PhysRevB.76.064120.

Lu, Haibao, Yanju Liu, Jihua Gou, Jinsong Leng, and Shanyi Du. 2010. "Electrical Properties and Shape-Memory Behavior of Self-Assembled Carbon Nanofiber Nanopaper Incorporated with Shape-Memory Polymer." *Smart Materials and Structures* 19 (7): 075021. https://doi.org/10.1088/0964-1726/19/7/075021.

Lu, Jian Ping. 1997. "Elastic Properties of Carbon Nanotubes and Nanoropes." *Physical Review Letters* 79 (7): 1297–1300. https://doi.org/10.1103/PhysRevLett.79.1297.

Manders and Bader. 1981. "The strength of hybrid glass/carbon fibre composites." *Journal of Materials Science* 16: 2233–2245. https://doi.org/10.1007/BF00542386

Mehranpour, H., M. Askari, M. Sasani Ghamsari, and H. Farzalibeik. 2010. "Study on the Phase Transformation Kinetics of Sol-Gel Drived TiO_2 Nanoparticles." *Journal of Nanomaterials* 2010 (January): 31:1–31:5. https://doi.org/10.1155/2010/626978.

Meng, Zhi-Ying, Li Chen, Hai-Yi Zhong, Rong Yang, Xiao-Feng Liu, and Yu-Zhong Wang. 2017. "Effect of Different Dimensional Carbon Nanoparticles on the Shape Memory Behavior of Thermotropic Liquid Crystalline Polymer." *Composites Science and Technology* 138 (January): 8–14. https://doi.org/10.1016/j.compscitech.2016.11.006.

Mo, Shang-Di, and W. Y. Ching. 1995. "Electronic and Optical Properties of Three Phases of Titanium Dioxide: Rutile, Anatase, and Brookite." *Physical Review B* 51 (19): 13023–32. https://doi.org/10.1103/PhysRevB.51.13023.

Mu, Tong, Liwu Liu, Xin Lan, Yanju Liu, and Jinsong Leng. 2018. "Shape Memory Polymers for Composites." *Composites Science and Technology* 160 (May): 169–98. https://doi.org/10.1016/j.compscitech.2018.03.018.

Naik, Gururaj V., Vladimir M. Shalaev, and Alexandra Boltasseva. 2013. "Alternative Plasmonic Materials: Beyond Gold and Silver." *Advanced Materials* 25 (24): 3264–94. https://doi.org/10.1002/adma.201205076.

Neumann, Oara, Alexander S. Urban, Jared Day, Surbhi Lal, Peter Nordlander, and Naomi J. Halas. 2013. "Solar Vapor Generation Enabled by Nanoparticles." *ACS Nano* 7 (1): 42–9. https://doi.org/10.1021/nn304948h.

Ohki, Takeru, Qing-Qing Ni, Norihito Ohsako, and Masaharu Iwamoto. 2004. "Mechanical and Shape Memory Behavior of Composites with Shape Memory Polymer." *Composites Part A: Applied Science and Manufacturing* 35 (9): 1065–73. https://doi.org/10.1016/j.compositesa.2004.03.001.

Okafor, Christian Emeka, Dominic Ugochukwu Okpe, Okwuchukwu Innocent Ani, and Ugochukwu Chuka Okonkwo. 2022. "Development of Carbonized Wood/Silicon Dioxide Composite Tailored for Single-Density Shoe Sole Manufacturing." *Materials Today Communications* 32 (August): 104184. https://doi.org/10.1016/j.mtcomm.2022.104184.

Okafor, Christian Emeka, Iweriolor Sunday, Okwuchukwu Innocent Ani, Nürettin Akçakale, Godspower Onyekachukwu Ekwueme, Peter Chukwuemeka Ugwu, Emmanuel Chukwudi Nwanna, and Anthony Chinweuba Onovo. 2023. "Biobased Hybrid Composite Design for Optimum Hardness and Wear Resistance." *Composites Part C: Open Access* 10 (March): 100338. https://doi.org/10.1016/j.jcomc.2022.100338.

Patel, Dinesh K., Arpan Biswas, and Pralay Maiti. 2016. "6- Nanoparticle-Induced Phenomena in Polyurethanes." In *Advances in Polyurethane Biomaterials*, edited by Stuart L. Cooper and Jianjun Guan, 171–94. Woodhead Publishing. https://doi.org/10.1016/B978-0-08-100614-6.00006-8.

Peijs, A.A.J.M., P. Catsman, L.E. Govaert, and P.J. Lemstra. 1990. "Hybrid composites based on polyethylene and carbon fibres Part 2: influence of composition and adhesion level of polyethylene fibres on mechanical properties." Composites 21 (6): 513–521. https://doi.org/10.1016/0010-4361(90)90424-U.

Pilate, Florence, Antoniya Toncheva, Philippe Dubois, and Jean-Marie Raquez. 2016. "Shape-Memory Polymers for Multiple Applications in the Materials World." *European Polymer Journal* 80 (July): 268–94. https://doi.org/10.1016/j.eurpolymj.2016.05.004.

Qi, Xiaodong, Hao Xiu, Yuan Wei, Yan Zhou, Yilan Guo, Rui Huang, Hongwei Bai, and Qiang Fu. 2017. "Enhanced Shape Memory Property of Polylactide/Thermoplastic Poly(Ether) Urethane Composites via Carbon Black Self-Networking Induced Co-Continuous Structure." *Composites Science and Technology* 139 (February): 8–16. https://doi.org/10.1016/j.compscitech.2016.12.007.

Qi, Xiaodong, Xuelin Yao, Sha Deng, Tiannan Zhou, and Qiang Fu. 2014. "Water-Induced Shape Memory Effect of Graphene Oxide Reinforced Polyvinyl Alcohol Nanocomposites." *Journal of Materials Chemistry A* 2 (7): 2240–9. https://doi.org/10.1039/C3TA14340F.

Rajkumar, Govindaraju, Jagannathan Srinivasan, and Latchupathi Suvitha. 2015. "Natural Protein Fiber Hybrid Composites: Effects of Fiber Content and Fiber Orientation on Mechanical, Thermal Conductivity and Water Absorption Properties." *Journal of Industrial Textiles* 44 (5): 709–24. https://doi.org/10.1177/1528083713512355.

Ratna, Debdatta, and J. Karger-Kocsis. 2008. "Recent Advances in Shape Memory Polymers and Composites: A Review." *Journal of Materials Science* 43 (1): 254–69. https://doi.org/10.1007/s10853-007-2176-7.

Rodriguez, Erika D., Xiaofan Luo, and Patrick T. Mather. 2011. "Linear/Network Poly(ε-Caprolactone) Blends Exhibiting Shape Memory Assisted Self-Healing (SMASH)." *ACS Applied Materials & Interfaces* 3 (2): 152–61. https://doi.org/10.1021/am101012c.

Safri, Syafiqah Nur Azrie, Mohamed Thariq Hameed Sultan, Mohammad Jawaid, and Kandasamy Jayakrishna. 2018. "Impact Behaviour of Hybrid Composites for Structural Applications: A Review." *Composites Part B: Engineering* 133 (January): 112–21. https://doi.org/10.1016/j.compositesb.2017.09.008.

Sokolowski, Witold M., and Seng C. Tan. 2007. "Advanced Self-Deployable Structures for Space Applications." *Journal of Spacecraft and Rockets* 44 (4): 750–4. https://doi.org/10.2514/1.22854.

Stankovich, Sasha, Dmitriy A. Dikin, Geoffrey H. B. Dommett, Kevin M. Kohlhaas, Eric J. Zimney, Eric A. Stach, Richard D. Piner, SonBinh T. Nguyen, and Rodney S. Ruoff. 2006. "Graphene-Based Composite Materials." *Nature* 442 (7100): 282–6. https://doi.org/10.1038/nature04969.

Swolfs, Yentl, Ignaas Verpoest and Larissa Gorbatikh. 2019. Recent advances in fibre-hybrid composites: materials selection, opportunities and applications. *International Materials Reviews* 64 (4): 181–215. doi: 10.1080/09506608.2018.1467365

Swolfs, Yentl, Larissa Gorbatikh, and Ignaas Verpoest. 2014. "Fibre Hybridisation in Polymer Composites: A Review." *Composites Part A: Applied Science and Manufacturing* 67 (December): 181–200. https://doi.org/10.1016/j.compositesa.2014.08.027.

Tiwari, Nilesh, and AbdulHafiz A. Shaikh. 2019. "Flexural Analysis of Thermally Actuated Fiber Reinforced Shape Memory Polymer Composite." *Advances in Materials Research* 8 (4): 337–59. https://doi.org/10.12989/amr.2019.8.4.337.

Tiwari, Nilesh, and AbdulHafiz A. Shaikh. 2021a. "Buckling and Vibration Analysis of Shape Memory Laminated Composite Beams under Axially Heterogeneous In-Plane Loads in the Glass Transition Temperature Region." *SN Applied Sciences* 3 (4): 1–15. https://doi.org/10.1007/s42452-021-04438-2.

Tiwari, Nilesh, and AbdulHafiz A. Shaikh. 2021b. "Micro Buckling of Carbon Fiber in Triple Shape Memory Polymer Composites under Bending in Glass Transition Regions." *Materials Today: Proceedings* 44 (6): 4744–8. https://doi.org/10.1016/j.matpr.2020.10.961.

Tiwari, Nilesh, and AbdulHafiz A. Shaikh. 2022a. "Effect of Graphene Nanoplatelets on the Thermomechanical Behaviour of Smart." In *Aerospace and Associated Technology*, edited by Anup Ghosh, Kalyan Prasad Sinhamahapatra, Ratan Joarder, and Sikha Hota, 300–4. Routledge.

Tiwari, Nilesh, and AbdulHafiz A. Shaikh. 2022b. "Influence of Carboxyl-Functionalized Graphene Nanoplatelets on the Thermomechanical and Morphological Behavior of Shape Memory Nanocomposites." *Composites: Mechanics, Computations, Applications: An International Journal* 13 (3): 4744–8. https://doi.org/10.1615/CompMechComputApplIntJ.2022043126.

Tiwari, Nilesh, and AbdulHafiz A. Shaikh. 2021c. "Micro-Buckling of Carbon Fibers in Shape Memory Polymer Composites under Bending in the Glass Transition Temperature Region." *Curved and Layered Structures* 8 (1): 96–108. https://doi.org/10.1515/cls-2021-0009.

Vita, Alessandro De, Jean-Christophe Charlier, X. Blase, and Roberto Car. 1999. "Electronic Structure at Carbon Nanotube Tips." *Applied Physics A: Materials Science and Processing* 68 (3): 283–6. https://doi.org/10.1007/s003390050889.

Wang, Zhong Lin, Ruiping.P Gao, P. Poncharal, Walt A. de Heer, Z. R. Dai, and Z. W. Pan. 2001. "Mechanical and Electrostatic Properties of Carbon Nanotubes and Nanowires." *Materials Science and* Engineering: C 16 (1): 3–10. https://doi.org/10.1016/S0928-4931(01)00293-4.

West, Paul R., Satoshi Ishii, Gururaj V. Naik, Naresh Kumar Emani, V. M. Shalaev, and Alexandra Boltasseva. 2010. "Searching for Better Plasmonic Materials." *Laser & Photonics Reviews* 4 (6): 795–808. https://doi.org/10.1002/lpor.200900055.

Wong, Eric W., Paul E. Sheehan, and Charles M. Lieber. 1997. "Nanobeam Mechanics: Elasticity, Strength, and Toughness of Nanorods and Nanotubes." *Science* 277 (5334): 1971–5. https://doi.org/10.1126/science.277.5334.1971.

Wu, Zhong-Shuai, Guangmin Zhou, Li-Chang Yin, Wencai Ren, Feng Li, and Hui-Ming Cheng. 2012. "Graphene/Metal Oxide Composite Electrode Materials for Energy Storage." *Nano Energy* 1 (1): 107–31. https://doi.org/10.1016/j.nanoen.2011.11.001.

Xu, Jia-Zhuang, Gan-Ji Zhong, Benjamin S. Hsiao, Qiang Fu, and Zhong-Ming Li. 2014. "Low-Dimensional Carbonaceous Nanofiller Induced Polymer Crystallization." *Progress in Polymer Science, Topical Issue on Composites* 39 (3): 555–93. https://doi.org/10.1016/j.progpolymsci.2013.06.005.

Young, Robert J., Ian A. Kinloch, Lei Gong, and Kostya S. Novoselov. 2012. "The Mechanics of Graphene Nanocomposites: A Review." *Composites Science and Technology* 72 (12): 1459–76. https://doi.org/10.1016/j.compscitech.2012.05.005.

Yu, Juhong, Hong Xia, Akira Teramoto, and Qing-Qing Ni. 2017. "Fabrication and Characterization of Shape Memory Polyurethane Porous Scaffold for Bone Tissue Engineering." *Journal of Biomedical Materials Research Part A* 105 (4): 1132–7. https://doi.org/10.1002/jbm.a.36009.

Yu, Min-Feng, Bradley S. Files, Sivaram Arepalli, and Rodney S. Ruoff. 2000. "Tensile Loading of Ropes of Single Wall Carbon Nanotubes and Their Mechanical Properties." *Physical Review Letters* 84 (24): 5552–5. https://doi.org/10.1103/PhysRevLett.84.5552.

Yu, Min-Feng, Oleg Lourie, Mark J. Dyer, Katerina Moloni, Thomas F. Kelly, and Rodney S. Ruoff. 2000. "Strength and Breaking Mechanism of Multiwalled Carbon Nanotubes Under Tensile Load." *Science* 287 (5453): 637–40. https://doi.org/10.1126/science.287.5453.637.

Zhang, Chun-Sheng, and Qing-Qing Ni. 2007. "Bending Behavior of Shape Memory Polymer Based Laminates." *Composite Structures* 78 (2): 153–61. https://doi.org/10.1016/j.compstruct.2005.08.029.

Zhang, Ruirui, Xiaogang Guo, Yanju Liu, and Jinsong Leng. 2014. "Theoretical Analysis and Experiments of a Space Deployable Truss Structure." *Composite Structures* 112 (June): 226–30. https://doi.org/10.1016/j.compstruct.2014.02.018.

3 Fabrication and Characterization

A recent method used for manufacturing fiber-reinforced composite components involved the use of a hand layup technique, which was reliable but slow and labor-intensive. However, with the growing interest in mass production of composite parts, manufacturing methods like pultrusion, compression molding, and filament winding have become more prevalent. These processes have been around for many years, but research into improvements in their characteristics and manufacturing process optimization began in the mid-1970s.

Another manufacturing technique is resin transfer molding (RTM), which has gained significant attention in various applications like the aerospace and automotive industries. RTM has the ability to produce complex-shaped composite structures with high production rates. With the addition of different techniques and automation, the manufacturing of fiber-reinforced polymer composites has reached a remarkable pace. Manufacturing techniques also use fast-curing resins, different upcoming fiber forms, and high-quality inspection techniques.

In recent years, the automotive industry has played an important role in driving the improvement of new manufacturing techniques for composite parts, with a focus on producing parts at a lower cost and at a faster rate. As a result, the field has seen increased research into innovative techniques like 3D printing and improved fiber placement techniques. These advances are helping to make composite materials more accessible and affordable for a wider range of applications. Overall, progress in manufacturing technology for fiber-reinforced polymer composites continues to evolve, driven by a combination of industrial needs and scientific advancements.

3.1 BASIC CONCEPTS

Composite construction can be achieved in many ways, regardless of its complexity, and each method has its own set of advantages and disadvantages. Several factors come into play when selecting a manufacturing method, including the type of matrix and fibers used, the geometry of the final product, and cost-effectiveness. Two critical factors that govern production processes are pressure and temperature. While pressure is important to allow the highly viscous resin to pass and bind the initially separate fibers, high temperatures are necessary for the chemical reaction of the resin to occur. The curing process refers to the chemical process that leads to crosslinking into the resin, and the duration of the curing cycle determines the time required for complete curing.

DOI: 10.1201/9781003391760-3

3.1.1 DEGREE OF CURE

Experimentally, the degree of recovery is assessed using a differential scanning calorimeter (DSC), as shown in Figure 3.1. Processing compounds will benefit from this
information.

The entire amount of heat required to finish a curing process (i.e., 100% degree
of cure) is equivalent to the portion underneath the curve between heat generation
rate and time produced in a dynamic heating trial. This is represented in Eq. (3.1) as:

$$H_R = \int_0^{t_f} \left(\frac{dQ}{dt}\right)_d dt \tag{3.1}$$

where

H_R = reaction heat

$\left(\dfrac{dQ}{dt}\right)_d$ = heat generation rate in a dynamic testing

t_f = time required to complete the reaction.

Degree of cure at any instance, t is defined by

$$\alpha_c = \frac{H}{H_R} \tag{3.2}$$

where
H = heat released in instance t and H_R = heat of reaction.

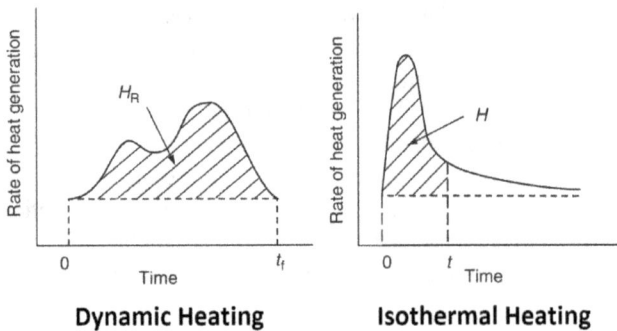

Dynamic Heating   **Isothermal Heating**

FIGURE 3.1 Illustration of the heat generation rate in a differential scanning calorimeter
(DSC) (Mallick 2007).

3.1.2 GEL TIME

When the curing time and temperature of a polymer are increased, the viscosity of the adhesive increases during the hardening process. At the beginning of the hardening process, the viscosity rises slowly. However, after a certain level of cure is reached, the density of the adhesive increases rapidly. This period is known as the gel time.

The exotherm curve, i.e., temperature–time curve, is used to measure gel time, as shown in Figure 3.2. Point A on the curve represents the time it takes to bring the catalyst–resin mixture to bath temperature, which signals the start of the curing process. As the curing process continues, the liquid mixture transforms into a gel-like substance, and the exothermic reaction generates heat, causing the temperature to rise. This rise in temperature accelerates the reaction rate as the catalyst decomposes more quickly. The temperature then increases rapidly to high values due to the heat production rate being greater than the heat loss rate to the surrounding area. The rate of heat production decreases as the curing process nears to completion, which leads to a temperature drop. The exothermic rise in temperature found in a gel-time measurement is influenced by the resin composition (degree of unsaturation) and the resin-to-catalyst ratio. The cure rate is determined by the slope of the exotherm graph, which primarily depends on the catalyst's reactivity.

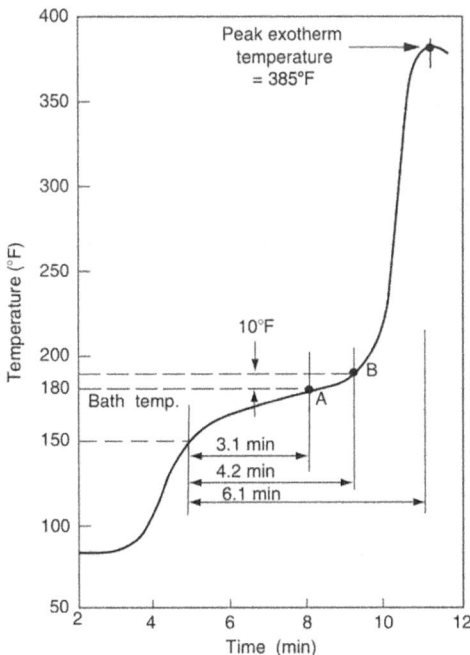

FIGURE 3.2   Temperature versus time curve to estimate gel time (Mallick 2007).

3.2   BAG-MOLDING PROCESS

This process is mostly used where production rate is not an important objective, like in the aerospace industry. Prepreg containing fibers in epoxy resin, which is partially cured, is considered an initial material requirement for bag-molding processes.

A prepreg typically has 42 wt% epoxy. The laminate that cured contained 50% of fibers if prepreg was permitted to dry without loss of resin. The real fiber content in the cured composite is 60 vol%, which is regarded as an industry standard for aircraft use because approximately 10 wt% of the resin oozes out at the time of molding process. Reduced void availability in the composite is achieved by removing confined air and leftover liquids with the extra resin streaming out of the prepreg. The current tendency is to use resin with a near-net composition, usually 34 wt%, and to allow only 1–2 wt% of resin loss during molding.

The design of a bag-molding method is shown in Figure 3.3. The plies of pre-preg are set up in the required order and at the desired fiber orientation angle on a Teflon-coated glass fabric separator that has been placed over the mold surface to avoid sticking. Using a cutting tool, which need not be more complicated than a mat knife, plies are sliced from prepreg sheet and trimmed in the required form, height, and orientation. The other methods include cutting dies, high-speed water sprays, and laser beams. Both manually and automatically operated tape-laying devices with numerical controls are capable of performing the layer-by-layer stacking procedure. On each ply, the reserve release film is peeled off prior to setting up the prepreg. To attach the prepreg to the fiber or the Teflon-coated glass cloth before it is laid up, a mild compaction pressure is used.

Once the layup process is completed, a permeable release fabric is placed over the stack of prepreg, along with a few pieces of bleeder paper. This allows any excess resin from the prepreg to seep out during the molding process, which is then collected by the bleeder sheets. To cover the entire layup, a Teflon-coated glass cloth separator layer is added, followed by a caul platter, a tinny vacuum bag that is heat-resistant, and a caul plate that is sealed with sealant. The autoclave is utilized to consolidate as well as solidify individual plies into a cohesive laminate by applying a combined effect of vacuum, external pressure, and heat to the entire assembly. While pressure

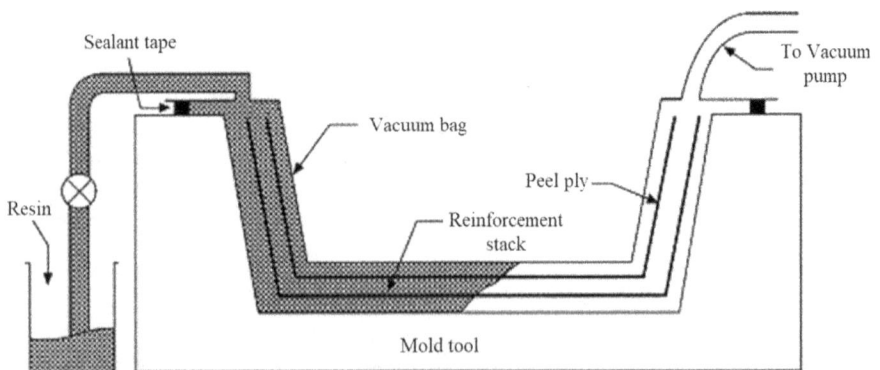

FIGURE 3.3   Schematic diagram of a bag-molding process (Mouritz 2012).

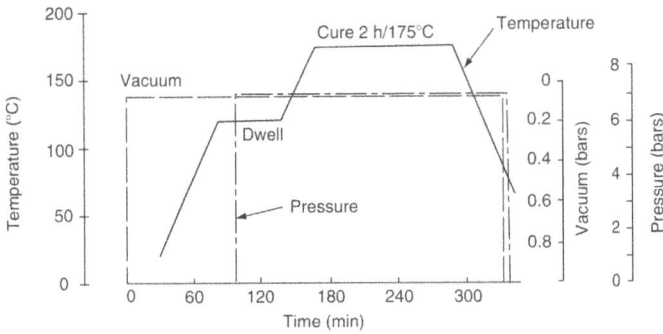

FIGURE 3.4 Representative curing cycle of a carbon fiber-reinforced epoxy prepag (Mallick 2007).

is essential for fusing the separate layers of a laminate, and is employed to eliminate any air or volatiles present.

The viscosity of resin in the B-staged prepreg plies initially declines when prepreg is heated with an autoclave, reaches a lesser degree, and then escalates (gels) as the preparation reaction starts and proceeds toward completion. The two-stage fix cycle for a carbon fiber-epoxy prepreg is shown in Figure 3.4. The first phase of this cure cycle involves gradually raising the temperature (2°C/min) up to 130°C and maintaining it there for almost 1 hour once the minimal resin viscosity is achieved.

During the temperature movement, an exterior weight is put on the prepreg, which causes an excess amount of resin to drip inside the bleeder sheets. This movement of resin is crucial as it allows for the release of trapped air as well as volatiles in the prepreg, ultimately reducing the number of voids inside the laminate after it is cured. The temperature of the autoclave is then raised to the resin's curing temperature, and pressure is maintained for at least 2 hours to ensure a predetermined amount of cure has taken place. The temperature is gradually decreased, and the laminate remains under pressure at the end of the curing period. Once the curing is complete, the laminates are removed and may be postcured at a high temperature inside an air-circulating furnace if necessary. The excess amount of resin available with the prepregs is critical for minimizing voids in the composite after curing. Face bleeding, which involves the resin flowing perpendicularly to the upper surface of the laminate, is favored over edge exploiting in bag-molding methods used to create thin shell or plate structures. This is because the width direction has a shorter resin-flow route prior to gelation than the border dimensions, making face bleeding more efficient. However, eliminating trapped air as well as volatiles from the center regions of laminate using the edge bleeding technique can be challenging, as the resin-flow route is comparatively lengthy in the direction of the edge.

3.3 COMPRESSION MOLDING

Sheet-molding composites (SMC) are formed into final goods using compression molding and matched templates. The main benefit of compression molding is its capacity to quickly create components with complex geometries. During the compression-molding process, features such as ribs, non-uniform thickness, flanges, bosses, apertures, and shoulders can be integrated. As a result, it facilitates elimination of a number of additional refining processes like drilling, forming, and soldering. Automation is an option for the complete molding procedure, including mold preparation, the addition of SMC to the mold, and component extraction from the mold. Therefore, the compression-molding method is appropriate for the mass manufacturing of composite components. Road rims, bumpers, and leaf springs are just a few of the structural car parts that are commonly made using this technique (Figure 3.5).

During the compression-molding process, a charge consisting of several rectangular plies is placed over the bottom half of a preheated mold chamber, covering 60%–70% of the surface. The mold is sealed quickly, and the upper half is gradually lowered to apply pressure to the charge until it reaches a preset level. The SMC material, which is inside mold fills the space and releases trapped air when pressure increases. The pressure used for molding varies between 1.4 and 34.5 MPa depending on the complexity of the component, flow duration, and fiber content, which command the SMC viscosity. High pressure is generally required to mold deep bosses and ribs. The mold is maintained at a temperature of 130°C–160°C. After the component has cured sufficiently under pressure, it is extracted from the mold using ejector pins.

In the mold chamber, complicated heat transmission and viscous flow occur during molding. The resin surface temperature quickly reaches the mold temperature and is consistent with the centerline temperature, as determined from temperature–time profiles noticed at different locations within the thick E-glass fiber-SMC molding. However, the poor thermal conductivity of the SMC material causes the centerline temperature to rise gradually until the curing reaction starts in the middle of the

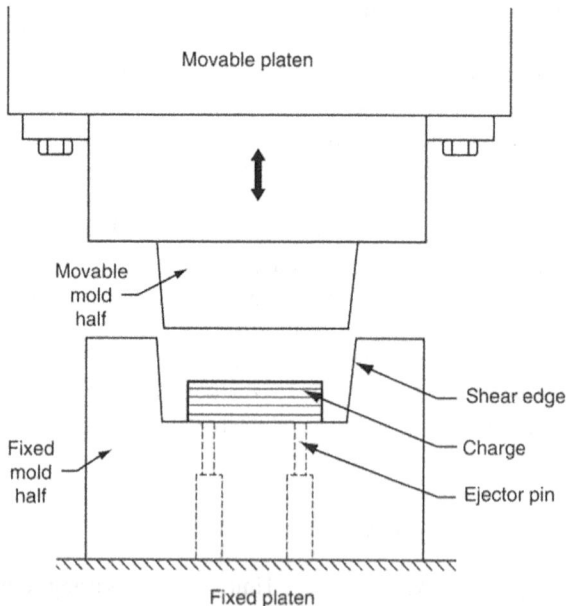

FIGURE 3.5   Illustration of a compression-molding process (Mallick 2007).

part's length. During the exothermic curing process, the heat is not effectively transmitted to the surface of the mold from the SMC charge inside which it is produced, resulting in a quick increment in the neutral line temperature to a maximum value. As the curing process approaches its conclusion, the midline temperature drops and reaches the mold surface temperature. The temperature increase is approximately constant across the thickness if the component is thin, and the material's highest temperature rarely surpasses the mold temperature.

Increasing the amount of filler in sheet-molding compound formulations causes a fall in the highest exotherm temperature. This is because the filler replaces some of the resin, which reduces the amount of heat released during curing. Additionally, the filler acts as a thermal sink, absorbing some of the heat. As the filler content increases, the time required for the material to reach its peak exotherm temperature decreases, which is correlated to the thickness of the component being cured.

To shorten the curing time in the mold, two effective methods are preheating the material to pregel temperatures before molding and using fast mold-closing rates during curing. Preheating can be done using dielectric warmers, which rapidly and evenly raise the temperature throughout the charge volume. This creates a consistent thermal gradient in the material, leading to uniform curing in the thickness direction. Using these techniques, the curing pressure in the molded component can be reduced, resulting in a shorter healing cycle.

The network structure produced by the hardening reaction with particulate fillers (such as MgO) breaks down as the temperature of the SMC resin rises in the mold. This causes reduction in the resin's viscosity. The material's ability to move through the mold is seriously constrained if it does not reach a low viscosity before gelling. Premature gelation, which takes place before the mold is fully filled, results in an unfinished component that may have interlaminar cracks and voids. The fundamental flow behavior of random fiber SMC with multicolored stacks in flat plaque mold holes has been researched by several researchers [12–14]. When a mold closes quickly, the sections extend uniformly (plug flow), but sliding occurs at the mold surface. This flow pattern is unaffected by charge thickness at rapid mold closure rates. On the other hand, at sluggish mold-closing rates, the charge thickness has a significant impact on the SMC flow pattern.

3.4 PULTRUSION

A continuous molding technique called pultrusion is used to create lengthy, straight structural components with consistent cross-sectional areas. Solid rods, empty tubes, smooth sheets, and beams with a range of cross sections, channels, and hat sections, including angles and wide-flanged sections, are some of the frequently produced pultruded products. Additionally, curved members and methods for creating varying cross sections along the length using pultrusion have been developed in recent times. Continuous-strand rovings that are horizontally aligned make up the majority of pultruded products. To increase its transverse strength, several layers of braided rovings or mats are added at or close to the exterior surface. A pultruded member's total fiber content may reach 70% by weight, but because mats or woven rovings are present, the lengthwise strength as well as modulus are lower

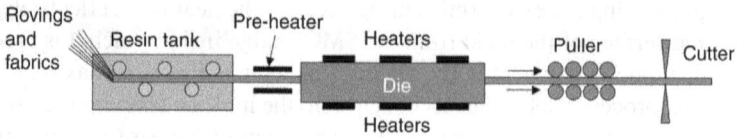

FIGURE 3.6   Illustration of the pultrusion process (Joshi 2012).

than those attained with all the same directional 0° fiber strands. Its mechanical characteristics are determined by the proportion of continuous thread rovings to mats or braided rovings.

Polyester and vinyl ester polymers are used as matrix substances in industrial applications. Epoxies have also been used, but they take longer to dry and are more difficult to remove from the die. The pultrusion method has also been applied to thermoplastic plastics like polysulfone and PEEK.

Figure 3.6 displays a standard pultrusion line, which involves pulling continuous filament rovings and carpets inside a resin tub that contains liquid resin, hardener, and additional components such as UV stabilizer, pigment, and fire resistance. The viscosity of the molten resin, mechanical process on the fibers and residence duration are altered to ensure that the fibers are fully saturated by resin. Thermoplastic polyester surfacing layers are injected into the fiber-resin stream to enhance the surface uniformity of the final molded product. The material is preheated and shaped using a lengthy die with a progressively decreasing portion along its length. To avoid early gelling, the entry region of the die is cooled with water while the remaining portion is carefully heated using oil warmers or electric heaters. The curing process can be accelerated by infrared radiation. The cured pultruded component is removed from the die using lifting rolls or blocks, and the die temperature, drawing speed, and length are regulated to ensure full curing before the component leaves the die. Finally, a diamond-impregnated saw is used to cut the material into appropriate lengths after chilling with air or water.

3.5   FILAMENT WINDING

Axisymmetric hollow components are created by wrapping a continuous band of rovings or monofilaments coated with resin around a revolving mandrel. Automotive drive shafts, oxygen tanks, chopper blades, pipes, conical rocket engine cases, and sizable subterranean gasoline storage tanks are a few examples of what the component filament winding is used for. Prepreg sheets and continuous fiber-reinforced sheet-molding materials are also produced using the filament-winding method. Slitting the coiled structure parallel to the mandrel plane creates the sheet.

A diagram of a fundamental filament-winding procedure is shown in Figure 3.7. A large quantity of fiber rovings is drawn from a set of creels in the form of a liquid resin solution that also contains a catalyst, liquid resin, colors, and UV absorbers. The fiber guides, also known as scissor bars, which are situated between each creel and the resin solution, are used to regulate fiber strain. The rovings are typically collected into a band just before entering the epoxy solution by running them through a stainless steel comb or a textile thread board.

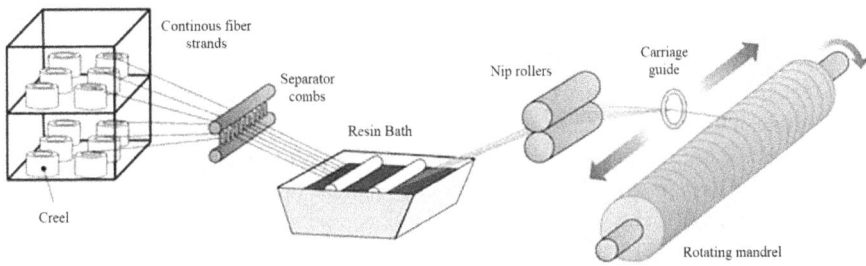

FIGURE 3.7 Illustration of the filament-winding process (Shrigandhi and Kothavale 2021).

The resin-impregnated rovings are drawn by a wiping mechanism at the end of the resin tank, which eliminates excess resin and regulates the thickness of the resin covering each roving. The most popular wiping tool is a pair of squeeze rollers where the upper roller's location can be changed to regulate the tension and resin content of fiber rovings. Pulling each roving through an opening individually, much like how wire is drawn, is another method for wiping the resin-impregnated rovings. This latter method offers greater resin composition management. However, it becomes challenging to rethread the fractured roving line by using its opening when a fiber breaks during a filament-winding operation.

The rovings are assembled into a flat band and placed on the frame after being carefully impregnated and cleaned. A straight rod, a ring, or a comb can be used to create a band. Like a tool stock on a lathe machine, the band former is typically mounted on a frame that moves parallel to the tube either back or forth. To produce the desired winding angle designs, the mandrel's winding speed and the carriage's traversing speed are both regulated. The typical turning pace is between 300 and 360 linear feet per minute (90–110 linear m/min). Though, slower rates are advised for winding for greater accuracy.

The helical winding process is the name given to the fundamental filament-winding procedure that produces a spiral winding design. The wind angle is the orientation of the moving band with regard to the mandrel plane. Any wind angle ranging from 0° (which corresponds to longitudinal winding) to 90° (which corresponds to hoop winding) can be achieved by changing the mandrel's speed and carriage feed rate. Fiber bands overlap at plus–minus the wind angle as the feed carriage travels back and forth, producing a weaving or interlocking effect. Another method for creating helical windings is to maintain the feed carriage still while moving the revolving mandrel back and forth. The mechanical characteristics of the helically coiled component are significantly influenced by the wind angle. The carriage moves about the longitudinal plane of a stationary (yet indexable) mandrel in polar winding, another form of filament-winding procedure. The mandrel is indexed to progress one fiber bandwidth after every carriage turn. As a result, there are no fiber crossings, and the fiber bands are close to one another. Two plies aligned at a plus–minus wind angle on two edges of the mandrel make up a full wrap.

3.5.1 WINDING STRATEGIES

The mechanical properties of the component are considerably affected by the winding arrangement (Rousseau, Perreux, and Verdière 1999). It is necessary to choose appropriate winding designs to achieve the minimum layup thickness. In filament winding, there are typically three kinds of winding patterns: helix, polar, and hoop (Figure 3.8) (Quanjin et al. 2017). The primary feature that sets one form of winding apart from another is the winding angle.

Helical Winding: Either helical or longitudinal winding involves winding filaments in the form of a helical pattern in the same direction at a specific winding angle (α), turning at the extreme end of the mandrel, and returning in the reverse direction at the same angle (α). The winding angle typically ranges from 5 to 80°; this pattern repeats itself until the desired thickness is achieved. This winding technique is useful for winding around pins, corners, or polar openings and can be applied to a variety of geometries. However, it requires specialized and more advanced winding equipment.

Hoop winding: Hoop winding, also called circumferential winding, involves wrapping strands approximately at right angles to the mandrel axis, or with a winding angle close to 90°. This winding method can be a specific type of spiral wrapping. Hoop winding is often used in combination with other different winding techniques to endure circumferential tension and is typically applied to circular portions of a structure.

Polar winding: During this process, the input fiber eye is rotated around its longitudinal axis at specific angles between 0 and 5°. The polar holes on the mandrel are positioned approximately perpendicular to the fiber. The length of the component and the distance between polar openings determine the extent of polar winding that can be achieved. This method is commonly used for spinning cylinders with a length-to-diameter (L/D) ratio of less than two.

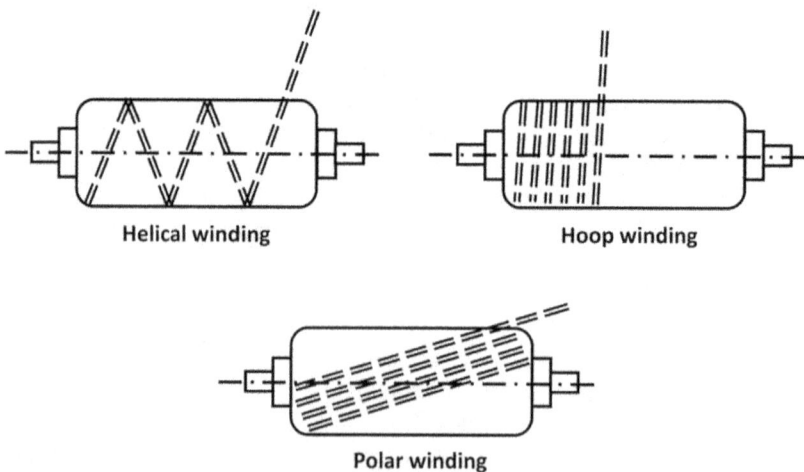

Helical winding   **Hoop winding**

Polar winding

FIGURE 3.8 Different winding strategies in filament winding (Sofi, Neunkirchen, and Schledjewski 2018).

3.6   RESIN TRANSFER MOLDING

RTM involves filling a two-part cast with woven roving, dry continuous-strand mat, or fabric and then sealing it, as shown in Figure 3.9. Liquid resin is catalyzed and poured into the mold through a sprue located at the lowest part of the casting chamber. The resin spreads throughout the mold, filling the dry fiber preform's area and covering the fibers while releasing air through the mold's air vents. Infusion pressure typically ranges from 69 to 690 kPa (10 to 100 psi). The cured part can be trimmed to meet precise measurements after it's detached from the mold.

The RTM process is able to begin with a prototype that already has the specified product's structure rather than flat-reinforcing layers like a continuous-strand mat. Using a preform can eliminate the need for the labor-intensive trimming process and improve moldability, particularly with complex forms or deep draws. Curing can be accomplished either in an air-circulating furnace at high temperatures or at room temperature, depending on the resin-catalyst system used.

The fundamental RTM procedure that was previously explained has many variants. In one of the variants, referred to as vacuum-assisted RTM, the liquid resin is drawn into the preform using suction, along with the resin infusion system, as shown in Figure 3.10. SCRIMP, or Seemann's Composite Resin Infusion Molding Process, is an additional RTM variant. In SCRIMP, vacuum is also used to draw liquid resin

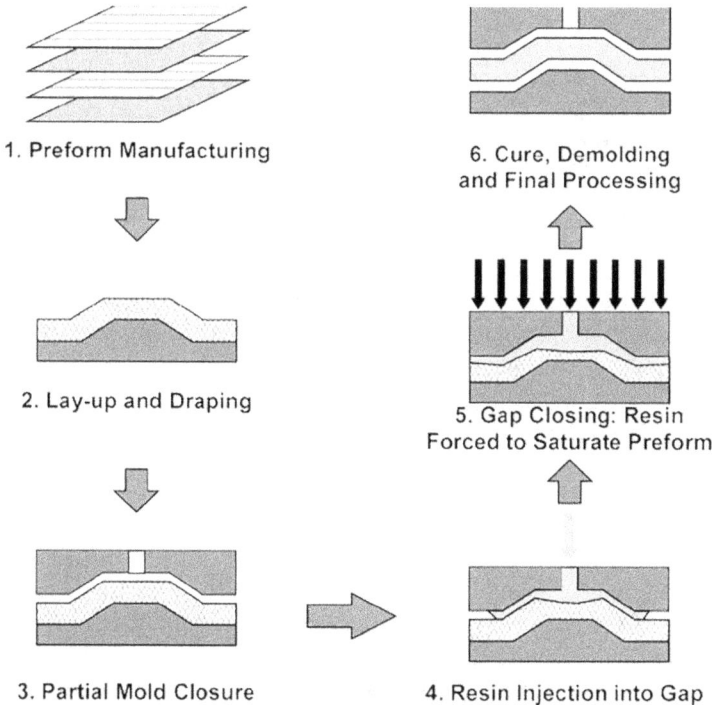

1. Preform Manufacturing

2. Lay-up and Draping

3. Partial Mold Closure

4. Resin Injection into Gap

5. Gap Closing: Resin Forced to Saturate Preform

6. Cure, Demolding and Final Processing

FIGURE 3.9   Illustration of the resin transfer molding process (Bhat et al. 2009).

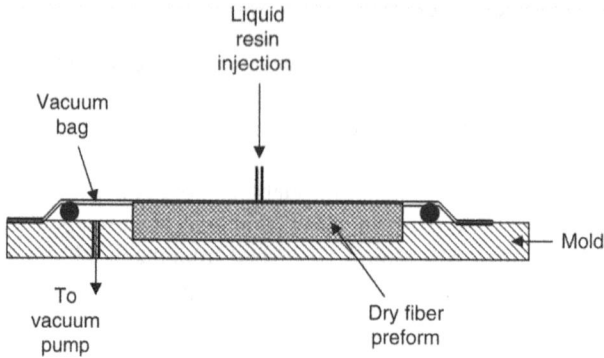

FIGURE 3.10   Illustration of vacuum-assisted resin transfer molding (Mallick 2007).

into the dry fiber mold, but to ensure that the resin is distributed evenly throughout the preform; a porous coating is applied to the preform. In order to give the liquid epoxy a simple flow route to the preform, the porous layer is chosen to have a very low flow resistance. One-sided rigid molds are utilized in vacuum-assisted RTM and SCRIMP. In a similar way to the bag-molding procedure discussed earlier, the preform is put on the hard mold surface and sealed by a vacuum bag.

There are different methods to create preforms, one of which is the spray-up technique. In this method, continuous fiber rovings are sprayed with cut strands of 12.7–76.2 mm (0.5–3 in.) length onto a pre-shaped screen. The fibers are held firmly on the screen by vacuum suction applied to the back surface of the screen. A binder is then sprayed onto the fibers to maintain the structure of the prepared material. Using a basic press and a pre-shaped die, continuous-strand carpets with random strands can be preformed. Binders are available in both thermoplastic and thermoset varieties, which can keep the established structure after pressing.

Another technique, the "cut and sew" method, involves cutting out different designs from woven fabric with bidirectional strands and stitching them together with polyester, glass, or Kevlar sewing threads. This method is useful for creating parts from woven fabric. Preforms can also be created in two or three dimensions using weaving and braiding techniques. Braiding is particularly effective for producing cylindrical preforms.

RTM has a much lower manufacturing cost and less complicated mold clamping needs than compression molding. In some instances, the two mold halves can be held together by a ratchet fastener or a set of screws and bolts. Parts can be resin transfer molded on low-tonnage machines because RTM is a low-pressure procedure. The potential to incorporate stiffeners, metal inserts, washers, and other materials within the molded composite is an additional benefit of the RTM process. Another option is to sandwich a foam core, which acts as a hollow component, between the top and bottom preforms. This gives the structure more rigidity and enables the molding of intricate three-dimensional structures in a single piece. Molding components like chairs, cabinet walls, hoppers, water containers, baths, and boat hulls have all been done effectively using the RTM method. Additionally, it provides a less expensive option to the capital- and labor-intensive bag-molding and compression-molding processes. It is best suited for manufacturing low- to mid-volume components, such as 5,000–50,000 parts annually.

3.7 CHARACTERIZATION

In controlled laboratory settings, mechanical and physical studies are typically used to identify material properties. Because fiber-reinforced composites are orthotropic, standard test procedures have been created that are often dissimilar from those used for conventional isotropic materials. Many of the properties addressed in this chapter are explored in connection with these special test techniques and their limitations.

3.7.1 TENSILE TESTING

The tensile test is conducted in accordance with ASTM D3039 using a straight-sided specimen with a consistent cross-sectional area. The specimen has beveled tabs attached at each end using adhesive bonding, as shown in Figure 3.11. To ensure the gage section experiences tensile failure, strain-compatible and compliant material is likely to be used for the end tabs, reducing stresses in the gripped area. Nonwoven E-glass-epoxy-balanced [0/90] cross-ply tabs have been found to produce satisfactory results. To mount the end tabs to the test component, any adhesive system that creates high elongation can be used.

Wedge action grips are used to hold the tensile material in the measuring apparatus while it is being tugged at the suggested cross-head movement rate of 2 mm/min (0.08 in/min). Electrical resistance strain gauges that are bound in the gauge portion of the specimen are used to measure longitudinal and transverse stresses. The stress test results of eight unidirectional laminates are used to calculate the longitudinal tensile modulus (E_{11}) and the main Poisson's ratio (v_{12}). The stress test results of 90° unidirectional laminates are used to calculate the transverse elasticity (E_{22}) and the Poisson's ratio (v_{21}).

A composite laminate's inhomogeneity and the statistical character of its component qualities frequently cause a significant variance in its tensile strength. The average strength (σ_{ave}), standard deviation (d) and coefficient of variance (CoV) are typically stated by Eqs. 3.3–3.5, assuming a normal distribution.

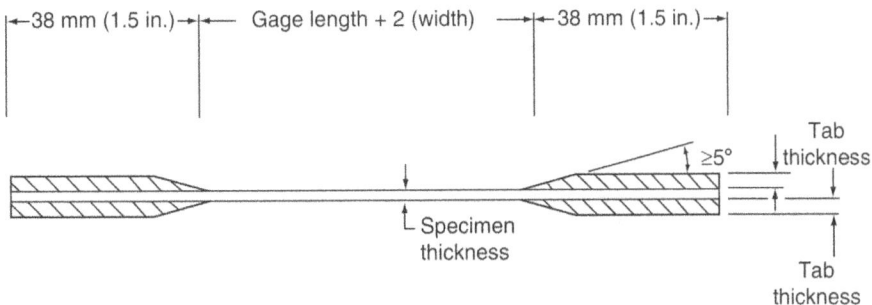

FIGURE 3.11 Specimen for tensile testing.

$$\sigma_{ave} = \sum \frac{\sigma_i}{n} \tag{3.3}$$

$$d = \sqrt{\frac{\sum(\sigma_i - \sigma_{ave})^2}{(n-1)}} \tag{3.4}$$

$$CoV = \frac{100d}{\sigma_{ave}} \tag{3.5}$$

where,

n = number of samples tested and σ_i = tensile strength of the ith sample.

Same directional structural laminates that contain fibers uniaxial to the loading path, which is tensile in nature, have a linear stress–strain curve until they fail, as shown in Figure 3.12. The specimens fail due to the rupture of fibers acted upon by tensile force, which is followed or attended by longitudinal splitting parallel to the fibers. This results in the failed area of 0° specimens appearing like a broom. However, off-axis specimens with angles between 0° and 90° may exhibit nonlinearity in the stress–strain curve. A tensile failure of the matrix or fiber–matrix interface causes ultimate failure for 90° specimens, where the fibers are transversal to the loading axis (tensile in nature). Failure at intermediate angles results from a synergy of fiber–matrix shear failure, matrix shear failure, and matrix tensile rupture. Most of these off-axis

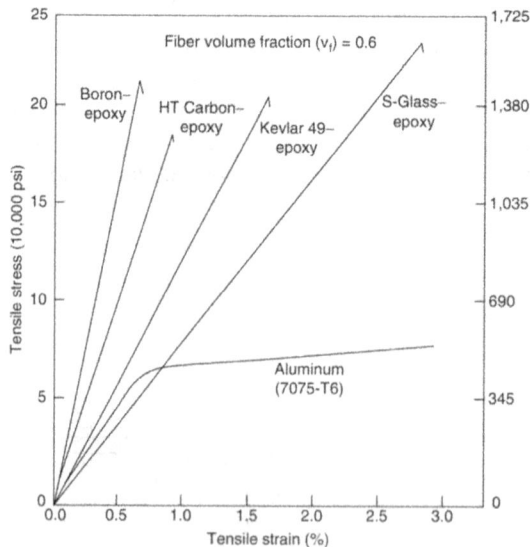

FIGURE 3.12 Tensile testing of the unidirectional laminates (Mallick 2007).

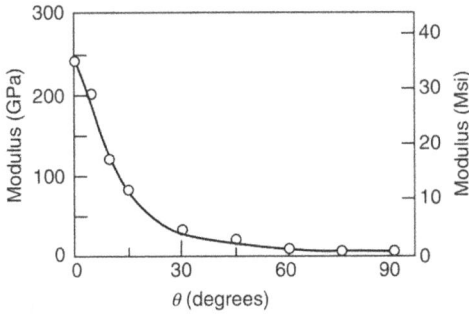

FIGURE 3.13 Relationship between tensile and modulus and fiber angle in carbon fiber-reinforced epoxy composites (Chamis and Sinclair 1978; Mallick 2007).

fiber-oriented specimens (Considering 90°) may have matrix craze marks throughout the gauge length parallel to the direction of fiber at low loads. The relation between the variation of the fiber angle and the respective tensile modulus is shown in Figure 3.13.

A cross-ply $[0/90]_S$ laminate investigated in the 08 direction exhibits a slightly nonlinear tensile stress–strain curve. But it is more commonly referred to as a bilinear curve (Figure 3.13) because the point where the 90° plies break is at the knee of the curve. At the fracture strain of 0° plies, laminates break completely.

To identify the variation in slope of the stress–strain plot at a defined knee point, the assumption is made that all 90° plies have failed at the knee point, i.e., they will not contribute to the laminate modulus. E_{11} and E_{22} denotes the moduli of 0° and 90° plies, respectively. The primary modulus of a cross-ply laminate can be approximated as in Eq. 3.6.

$$E = \frac{A_0}{A} E_{11} + \frac{A_{90}}{A} E_{22} \qquad (3.6)$$

where, A_0 and A_{90} = volume fractions of the 0° and 90° plies, respectively and $A = A_0 + A_{90}$. Tensile properties of the carbon fiber-reinforced epoxy composites with different laminate configurations are compiled in Table 3.1.

3.7.2 COMPRESSIVE TESTING

Due to material sideways buckling, it is challenging to quantify the compressive characteristics of thin composite laminates. To solve the buckling issue, various test procedures and model designs have been created (Whitney, Daniel, and Pipes 1982). The following is a description of three of these evaluation procedures.

Celanese test: It was one of the pioneer ASTM standard tests created for evaluating fiber-reinforced materials in compression, but now it is not a standard test due to a number of flaws. It uses a specimen with a straight side, tabs joined at the ends, and 10° grips that are tapered like collets and slide into casings with an interior taper that matches (Figure 3.14). In order to facilitate alignment and construction, an outer cylindrical casing

TABLE 3.1

Tensile Properties of the Carbon Fiber-Reinforced Epoxy Composites with Different Laminate Configurations (Freeman and Kuebeler 1974)

Laminate Configuration	Ultimate tensile strenth (UTS) (MPa)	Estimated First Ply Failure Stress (MPa)	Tensile Modulus (GPa)	Initial Tensile Strain (%)
[0]s	1378	–	151.6	0.3
[90]s	41.3	–	8.96	0.5–0.9
[±45]s	137.8	89.6	17.2	1.5–4.5
[0/90]s	447.8	413.4	82.7	0.5–0.9
[0_2/±45]s	599.4	592.5	82.7	0.8–0.9
[0/±60]s	461.6	323.8	62	0.8–0.9
[0/90/±45]s	385.8	275.6	55.1	0.8–0.9

is used. The grip on the specimen tightens as the compressive pressure is applied at the extremities of the tapered sleeves, and the frictional forces transferred through the end tabs squeeze the specimen's gauge section. In the gauge portion, strain gauges are mounted to record longitudinal and transverse strain measurements that are used to calculate compressive modulus and Poisson's ratio.

IITRI test: The compression test known as the Illinois Institute of Technology Research Institute (IITRI) trial was first established at the Illinois Institute of Technology Research Institute. Subsequently, it was implemented by

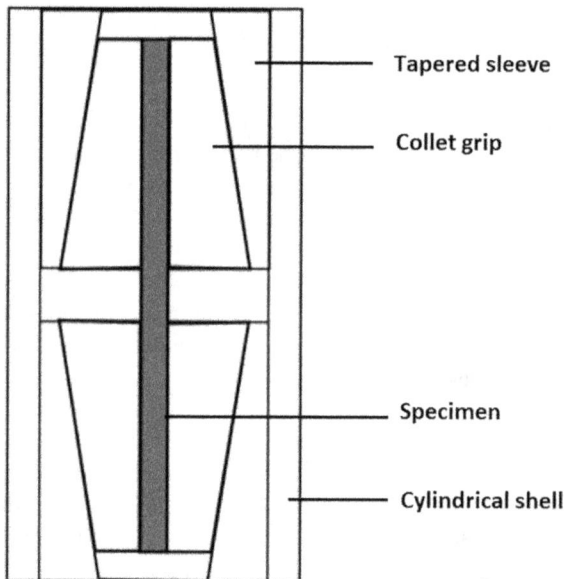

FIGURE 3.14   Attachment for the Celanese test.

ASTM D3410 as the preferred compression examination for fiber-reinforced composites. Compared to the Celanese test, it differs in that it replaces conical wedge grips (Figure 3.15) and uses flat wedge grips, which provide better contact between the collet and wedge, resulting in improved axial alignment. Additionally, flat wedge grips provide compensation for variations in sample depth. During testing, the IITRI test fixture has two parallel guide pins that are set in its bottom half. These pins then move into two roller bushings that are located in the fixture's top half, which helps to keep the two halves in proper lateral alignment with one another. According to the specifications for the typical sample, its length should be 140 μm, with the middle 12.7 μm acting as the gauge length while remaining undisturbed. Tabbing is preferred overusing untabbed specimens since it helps prevent surface damage and ends crushing of the test specimen if the force applied for clamping becomes too high.

Sandwich edgewise compression test: For the purpose of ensuring lateral stability during a compressive load test, these specimens for the compression testing consist of two flat-sided pieces that have been bonded to an aluminum honeycomb core. The end caps are utilized to support the specimen and prevent end crushing. For composite laminates, the resulting average compressive stress is designed with the assumption that the core bears no load. Table 3.2 presents a range of compressive properties for boron fiber-epoxy and carbon fiber-epoxy laminates tested using a sandwich edgewise compression method. The data from the experiment indicate that the compressive properties of the composite material are expressively exaggerated by both the type of fiber used and the configuration of the laminate.

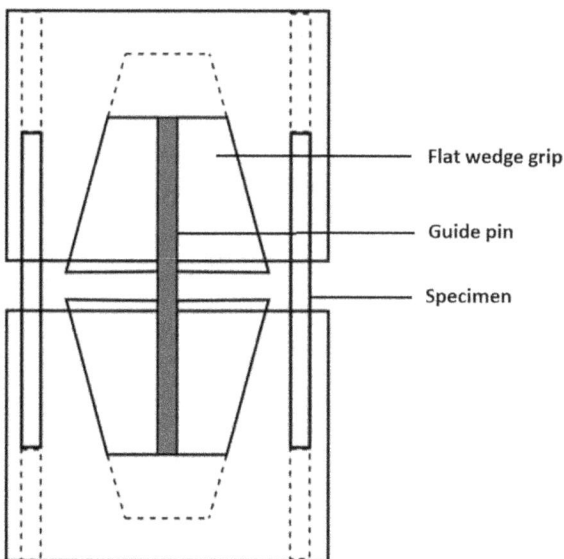

Flat wedge grip

Guide pin

Specimen

FIGURE 3.15 Attachment for the IITRI test.

TABLE 3.2

The Charateristics of the Carbon Fiber-Reinforced Epoxy Composites with Different Laminate Configurations (Weller 1977)

Laminate Configuration	Compressive Strength (MPa)	Compressive Modulus (GPa)
[0]	1219.5	110.9
[90]	194.3	13.1
[±45]	259.7	15.6
[0/90]	778.6	60.6
[0/±45/90]	642.8	46.4

3.7.3 Flexural Testing

By using ASTM test standard D790, flexural characteristics like flexural strength and modulus are evaluated. A composite beam specimen with a rectangular cross section is loaded in one of two bending modes: three points or four points (Figure 3.16). Large span-thickness ratios ($L = h$) are advised for both modes. In this section, discussion will be on the three-point bending test.

The flexural strength of a substance is defined as the highest fiber stress on the tension side during failure of a flexural specimen. As a result, the flexural strength in a three-point flexural measurement is provided by using the homogeneous beam theory (E).

$$\sigma_{UF} = \frac{3P_{max}L}{2bh^2} \tag{3.7}$$

where, P_{max} = highest failure load; b = model width; h = sample thickness; and L = span between the two supports.

The load-deflection curve's initial slope (m) is used to determine the flexural modulus:

$$E_F = \frac{mL^3}{4bh^3} \tag{3.8}$$

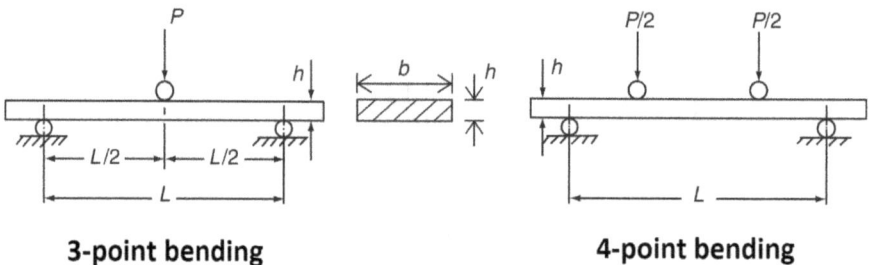

3-point bending   **4-point bending**

FIGURE 3.16 Flexural testing arrangement (Mallick 2007).

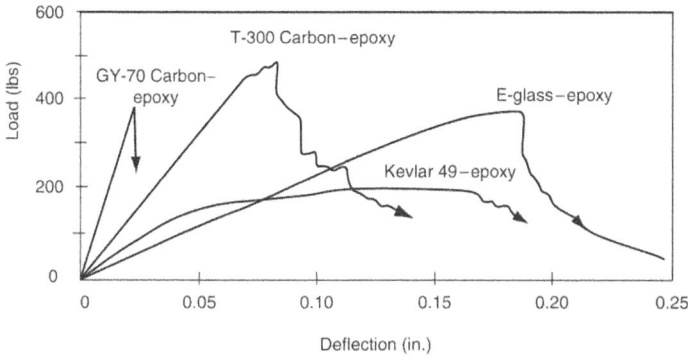

FIGURE 3.17 Comparison of the load-deflection curves in flexural testing for different fiber reinforcements (L/h = 11–16 and $\theta = 0°$) (Mallick 2007).

Figure 3.17 displays the load-deflection diagrams for four unidirectional 0° laminates: high-strength carbon (T-300), ultrahigh-modulus carbon (GY-70), E-glass fiber-reinforced epoxies, and Kevlar 49. The difference in slope of the load-deflection curves is due to variations in the respective fiber modulus of each material. While the GY-70 laminate reveals brittle behavior, the other laminates demonstrate progressive failure modes comprising debonding, fiber failure, and delamination. In the case of Kevlar 49, the load-deflection curve is highly nonlinear due to compressive yielding. Microbuckling of fibers is observed on the compression side of both T-300 and E-glass laminates. Therefore, it is recommended to use a great loading nose radius, as high contact stresses beneath the loading point can cause such damage.

The lamina stacking sequence crucially depends on flexural modulus and, consequently, does not associate with other moduli like tensile modulus, which is less dependent on stacking arrangement.

In angle-ply laminates, a bending moment causes both twisting and bending curvatures. Twisting curvature leads to lift supports in the diagonal corners of a flexural specimen, which affects the primary evaluated flexural modulus. The twisting curvature can be reduced by increasing the length-to-width ($L = b$) ratio and reducing the degree of orthotropy (i.e., decreasing $E_{11} = E_{22}$). Effects of some of the laminate configurations on the flexural properties of the composites are compiled in Table 3.3.

3.7.4 IN-PLANE SHEAR TESTING

Unidirectional fiber-reinforced composites characteristics of shear deformation in the plane, such as the modulus of shear G_{12} and the ultimate shear strength τ_{12u}, have been measured using a range of measurement techniques (Yeow and Brinson 1978). The following is a description of three popular in-plane stress test techniques for determining these two characteristics.

±45 **Shear test:** As prescribed in ASTM D3518, [+45/−45]s symmetric composite is subjected to uniaxial tension testing as part of the ±45 shear test. The tensile test standard ASTM D3039 specifications for specimen

TABLE 3.3

Flexural Properties of the Carbon Fiber-Reinforced Epoxy Composite with Different Laminate Configurations in 4-Point Bending Test (L/h = 32 and L/b = 4.8) (Whitney, Daniel, and Pipes 1982)

Laminate Configuration	Flexural Strength (MPa)	Flexural Modulus (GPa)
[0/±45/90]s	1219.5	68.9
[90/±45/0]s	141.2	18.6
[45/0/-45/90]s	263.9	47.54

measurements, test preparation, and test methodology are followed here. Eqs. 3.9 and 3.10 are used to draw a graph of the shear stress (τ_{12}) versus the shear strain (γ_{12}):

$$\tau_{12} = \frac{1}{2}\sigma_{xx} \tag{3.9}$$

$$\gamma_{12} = \varepsilon_{xx} - \varepsilon_{yy} \tag{3.10}$$

where, in the [+45/−45]s tensile specimen, σ_{xx}, ε_{xx}, and ε_{yy} represent tensile stress, in-plane strain, and transverse strain, respectively. Figure 3.18 illustrates the shear stress–shear strain diagram of a [±45]s boron–epoxy laminate.

Iosipescu shear test: Nicolai Iosipescu invented the Iosipescu shear test as per ASTM D5379, which was subsequently used by Walrath and Adams (1983) to evaluate the shear strength and elasticity of composite materials with fiber reinforcement. It utilizes a four-point bending fixture to evaluate a double V-notched test material (Figure 3.19). The gauge section of the specimen experiences a homogeneous transverse shear force, but in the notch plane, there is no bending moment at all. Numerous analyses have demonstrated that the notch plane is in a condition of pure shear, with the exception of the immediate region of the notch roots. When a notch is present, it causes a

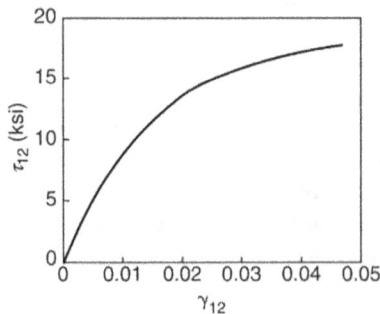

FIGURE 3.18 Comparison of the shear stress as a function of the shear strain for the quasi-static experiments (De Baere, Van Paepegem and Degrieck 2008).

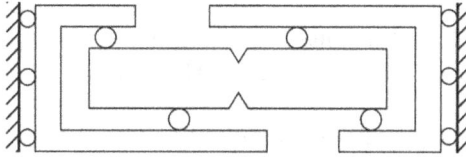

FIGURE 3.19 Test fixture for the Iosipescu shear test.

concentration of shear stress at the notch root. This shear stress concentration decreases with increasing notch angle and notch root radius but rises with increasing orthotropy, or when $E_{11} = E_{22}$. The standard Iosipescu examples have a notch angle of 90°, a notch depth of 20% of the specimen breadth, and a 1.3 mm notch root radius. In an Iosipescu shear test, the shear force is computed by Eq. 3.11 as

$$\tau_{12} = \frac{P}{wh} \tag{3.11}$$

The shear strain at the midpoint between the cuts is measured using a 45° strain patch that is centerd in the specimen's gauge section. Shear tension is expressed in Eq. 3.12 as

$$\gamma_{12} = \varepsilon_{+45°} - \varepsilon_{-45°} \tag{3.12}$$

0° examples should be used to measure the shear strength τ_{12u} and shear modulus G_{12} of a material consisting of continuous fibers that are used in the composite, according to an ASTM round-robin test (Dw 1990). In-shear shear properties of different fiber-laminated epoxy composites are stated in Table 3.4.

3.7.5 INTERLAMINAR SHEAR STRENGTH TESTING

The term "shear" refers to a collection of forces that work in parallel but in opposing directions. When two layers are joined together, the resulting connection has sufficient shear strength, and this property is known as interlaminar shear strength

TABLE 3.4

In-Shear Shear Properties of Different Laminated Composites with a Volume Fraction of 60% (Mallick 2007)

Materials	Shear Strength (MPa)		Shear Modulus (GPa)	
	[0]	[±45]s	[0]	[±45]s
Boron–epoxy	62	530.5	4.82	54.4
Carbon–epoxy	62	454.7	4.48	37.9
Kevlar 49–epoxy	55.1	192.9	2,07	20.7
S-Glass–epoxy	55.1	241.1	5.51	15.1

(ILSS). The short-beam shear (SBS) (SBS) is used to measure the ILSS of fiber-reinforced composites, which is the shear strength between the laminate planes of the composites. The test's execution is detailed in standards including ASTM D2344, EN 2563, and EN ISO 14130.

A three-point flexural test is performed over the small-span specimen, whose dimensions are assumed to be ($L = h$). The test was conducted for shear lamina failure in an in-plane direction. To explain the SBS test, the maximum normal stress (σ_{xx}) and maximum shear stress (τ_{xz}) are stated in Eqs. 3.13 and 3.14, considering the homogeneous beam equations.

$$\sigma_{xx} = \frac{3PL}{2bh^2} \tag{3.13}$$

$$\tau_{xz} = \frac{3P}{2bh} \tag{3.14}$$

Eqs. 3.13 and 3.14 demonstrate that as the L/h ratio diminishes, the maximum normal stress in the beam also declines, while the determined shear stress (about the neutral plane) is unaffected. Thus, despite the fact that the typical stress is still rather low, for suitably low L/h ratios, the maximum shear stress in the beam will achieve the ILSS of the material. Because of this, the beam will shatter in the interlaminar shear mode, where it will divide in a horizontal plane between the laminae (Figure 3.20). It is suggested that L/h ratios in the range of 4.5–5.0 be used in SBS tests. However, evaluating a few examples at different L/h ratios is usually necessary for determining the optimal L/h ratio for interlaminar shear failure. For lower L/h ratios, tensile failure can occur on the top surface of the specimen, while compressive failure can occur on the bottom surface (Daniels, Harakas, and Jackson 1971). The ILSS for the specified ultimate load may be determined using Eq. 3.14. The SBS test is widely used for evaluating samples of materials and checking their quality.

The in-plane shear strength (τ_{xyU}) and the ILSS (τ_{xzU}) are not the same. Furthermore, the shear modulus of a substance should not be found using the SBS test. One of the most important failure modes, interlaminar shear failure, is well known in fiber-reinforced composite laminates, despite the SBS test's limitations. Laminates with a brittle matrix, such as epoxy resin, are tested for their ILSS using an SBS test. The shear forces that occur in every flexure test are exploited here. Shear stresses are

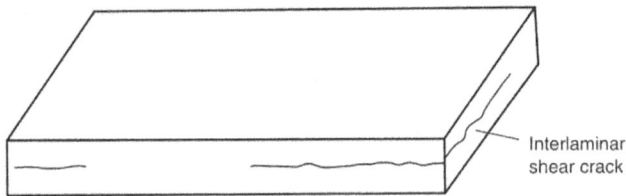

Interlaminar
shear crack

FIGURE 3.20 Schematic representation of interlaminar shear failure in unidirectional laminated composites.

much larger than normal stresses caused by a bending moment if the support span is tiny in relation to the thickness of the material. This allows the shear strength to be measured by creating a shear stress fracture in brittle matrix materials. The matrix transverse strength and matrix percentage amount are both increased to enhance the ILSS. Epoxies in general generate higher ILSS than vinyl ester and polyester resins in glass fiber-reinforced composites due to greater adhesion with glass fibers. The ILSS declines with rising void percentages, frequently linearly. The ILSS is also decreased by fabrication flaws like interior microcracks and desiccated strands.

3.7.6 FATIGUE TESTING

In many contexts, the material in question will be exposed to cyclic loading, and its fatigue properties come into play. This refers to the material's ability to withstand such loading. Under cyclic loads, the strength of a material is known to decrease significantly. Repeated cyclic loads or strains may cause brittle failure in even the most ductile metallic materials. Factors including stress, stress state, cycling mode, material composition, process history, and environmental circumstances all play a role in determining the failure cycle.

S–N behavior of composites is fundamental to the fatigue life of samples, and researchers use an S–N diagram, which depicts the association between the stress amplitude, or maximum stress, and the N to failure on a semilogarithmic scale. Researchers can obtain this figure by investigating several specimens at different stress levels under sinusoidal loading conditions.

Most of the fiber-reinforced composite material fatigue studies have been carried out using uniaxial tension–tension cycling (Figure 3.21). The utilization of tension-compression and compression–compression cycling is uncommon because of the risk of buckling failure in thin laminates. Flexural fatigue studies are able to produce completely reversed tension-compression cycling. A few interlaminar shear fatigue experiments and in-plane shear fatigue tests have also been carried out.

ASTM D3479 describes the method of tension–tension fatigue cycling testing, which involves using a sample with a straight edge and end tabs identical to those used in static tension tests. However, the heat generated by the internal damping of polymer matrix composites at high cycle frequencies can cause a noteworthy increase

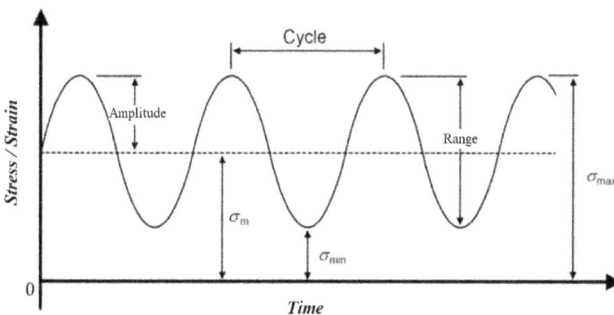

FIGURE 3.21   Stress–time curve in fatigue testing.

in the temperature of the object, negatively impacting its fatigue performance. To avoid this issue, it is preferred to conduct tests at low cyclic frequencies (<10 Hz). The tests are performed using stress-controlled and strain-controlled methods, where a persistent stress or strain value is maintained by cycling the material between predetermined maximum and minimum values. The material is subjected to cyclical maximum and minimum stresses in stress-controlled tests and cyclical maximum and minimum strains in strain-controlled testing.

Fiber-reinforced composite materials possess a unique characteristic where they gradually lose stiffness or become softer due to minuscule damages that occur long before any visible damage appears. This results in an increase in strain during load-controlled tests and a decrease in tension during strain-controlled tests. Microscopic damages also lead to a decline in the material's remaining strength. In order to determine the material's endurance, multiple stress tests are conducted until the specimen's stiffness or remaining strength falls below a predetermined threshold, rather than until the specimen completely fractures. Therefore, the failure cycle numbers may not always accurately indicate the specimen's lifespan until total fracture. Several researchers have attempted to plot the S-log N relationship as a straight line for various types of fiber-reinforced composites, as indicated in Eq. 3.15.

$$S = \sigma_U (m \log N + b) \qquad (3.15)$$

where,

S = maximum fatigue stress; N = failure cycle numbers; σ_U = average static strength; and m and b = constants.

Table 3.5 lists the m and b values for some of the epoxy matrix compounds. In this table, R denotes the ratio of minimum stress to maximum stress in a fatigue cycle.

S–N graphs can also be stated as Eq. 3.16:

$$\frac{S}{\sigma_U} N^d = c \qquad (3.16)$$

TABLE 3.5

Values of Constant in S–N Relation for Different Laminated Composites (Pankar K. Mallick 2007)

Material Configuration	Layup	R	m	b
E-glass–ductile epoxy	0°	0.1	−0.1573	1.3743
T-300 Carbon–ductile epoxy	0°	0.1	−0.0542	1.0420
E-glass–brittle epoxy	0°	0.1	−0.1110	1.0935
T-300 Carbon–brittle epoxy	0°	0.1	−0.0873	1.2103
E-glass–epoxy	[0/±45/90]s	0.1	−0.1201	1.1156
E-glass–epoxy	[0/90]s	0.05	−0.0815	0.934

where c and d are constants. Similar formulas can be used to write ε–N plots from fatigue experiments with regulated strain. The fatigue life, also known as the number of rounds to failure, typically shows a sizable amount of dispersion. The chance of fatigue life surpassing L is given by a two-parameter Weibull distribution and can be stated in Eq. 3.17 as

$$F(L) = \exp\left[-\left(\frac{L}{L_0}\right)^{\alpha_f} \right]$$

(3.17)

where α_f is the fatigue shape measure. The position measure for the distribution of fatigue life is L_0 (cycles).

When Hahn and Kim (1975) compared the data on static strength and fatigue life, they came up with Eq. 3.18 as a way to co-relate the static strength and fatigue data in a unidirectional $0°$ E-glass-epoxy composite. This equation was developed to co-relate the static strength and fatigue data. This association indicates that there is a proportionality between the sample's static strength and its fatigue life.

$$\frac{L}{L_0} = \left(\frac{S}{\sigma_U}\right)^{\alpha/\alpha_f}$$

(3.18)

3.7.6.1 Tension–Tension Fatigue

Consider the carbon fiber-reinforced thermoset polymer having fiber orientation along $0°$ and have S–N curves that remain almost horizontal and in the static scatter band during tension–tension stress tests. When compared to carbon fibers with a lower modulus, the fatigue effect is slightly more pronounced. Furthermore, composites made of unidirectional $0°$ boron and Kevlar 49 fibers display exceptional fatigue strength under tension–tension stress.

Several groups of researchers have compiled evidence demonstrating the fatigue resistance of SMC composites when subjected to tension–tension conditions (Mallick 1981; Pipes 1974). SMC composites that have a matrix of polyester or vinyl ester reinforced with E-glass fibers also do not display a fatigue limit. Instead, the fatigue behavior of these materials is influenced by the ratio of chopped fibers to continuous fibers in the laminate.

3.7.6.2 Flexural Fatigue

In general, the tension–tension fatigue demonstration of fiber-reinforced composites is superior to the bending fatigue performance. This is shown in Figure 3.22, where the slope of the bending S–N curve for high-modulus carbon fibers is higher than the slope of the tension–tension S–N curve. Due to the compression side of composites' vulnerability, flexure has reduced fatigue strength.

3.7.6.3 Interlaminar Shear Fatigue

Numerous researchers have examined the interlaminar shear (τ_{xz}) mode fatigue properties of fiber-reinforced composite materials (Pipes 1974; Makeev 2013; DeTeresa, Freeman, and Groves 2004). SBS examples were used in the interlaminar

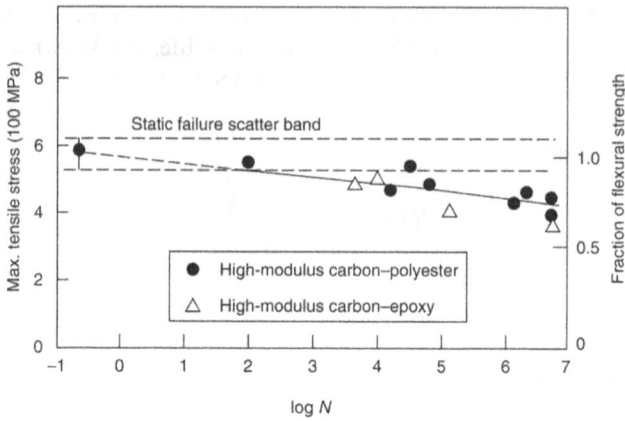

FIGURE 3.22 Flexural *S–N* diagram for 0° carbon fiber-reinforced polymer composites with $R = -1$ (Hahn and Kim 1976).

shear stress tests. Even though a unidirectional 0° carbon fiber-reinforced epoxy had a tension–tension fatigue strength close to 80% of its static tensile strength after 10^6 cycles, the interlaminar shear fatigue strength was only about 55% of that value (Figure 3.23). A unidirectional 0° boron–epoxy composite performed similarly to a unidirectional 0° carbon–epoxy composite in terms of interlaminar shear stress. To put things another way, a straight 0° S-glass-reinforced epoxy showed a contrary tendency. This material's interlaminar shear fatigue strength, in case you were wondering, was approximately 60% of its static ILSS, whereas the tension–tension fatigue strength was approximately 40% of its static tensile strength at 10^6 cycles. Contrary to the static interlaminar strengths, the high cycle interlaminar fatigue strength was not significantly influenced by fiber volume percentage or fiber surface treatment.

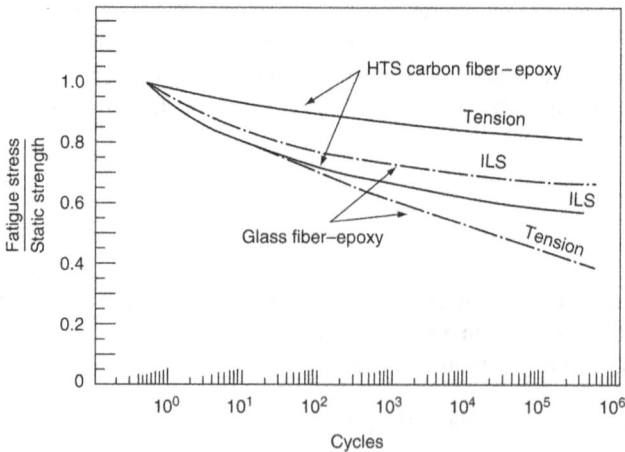

FIGURE 3.23 Interlaminar shear curves for unidirectional carbon and glass fiber-reinforced epoxy (Pipes 1974).

3.7.6.4   Torsional Fatigue

As shown in Figure 3.24, fatigue data under torsion has been presented for glass, Kevlar, carbon (HTS, High mod.) fiber-reinforced epoxy composite and compared with the solid rod specimen under shear strain loading. The result shows that with the application of shear cyclic repeating load over unidirectional composites, failure occurs approximately at 1,000 cycles and about half the static shear strain. Such rapid weakening results are not obtained quickly with short-beam interlaminar fatigue testing.

3.7.7   COEFFICIENT OF THERMAL EXPANSION

The coefficient of thermal expansion (CTE) measures how much a material expands or contracts in response to changes in temperature. Its number is used to calculate thermal stresses brought on by temperature fluctuations as well as dimensional variations.

Unreinforced plastics have a greater CTE than metals do. In general, a polymer's CTE decreases when reinforcements are added. The CTE of fiber-reinforced polymers can differ widely in values based on the fiber variety, alignment, and volume percentage. The longitudinal CTE (α_{11}) in unidirectional 0° laminates shows the properties of the material. Thus, in the longitudinal direction, carbon and Kevlar 49 fibers both generate a negative CTE, whereas glass and boron fibers create a positive CTE. The CTEs for unidirectional 0° laminates are distinct in in-plane and transverse orientations, similar to the situation with elastic properties (Table 3.6). Kevlar 49 fiber-reinforced epoxies show more heterogeneity in their CTE compared to carbon fiber-reinforced epoxies because Kevlar 49 has more asymmetry in its CTE.

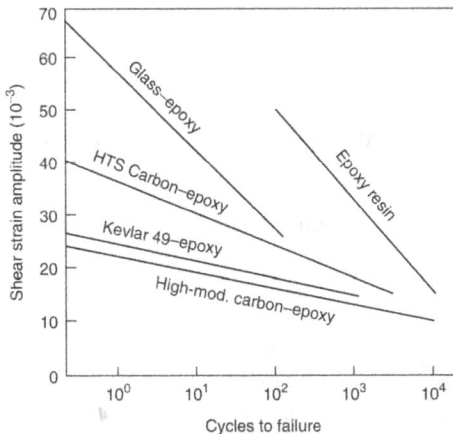

FIGURE 3.24   Torsional fatigue behavior of various unidirectional fiber-reinforced epoxy (Phillips and Scott 1977).

TABLE 3.6
Coefficient of Thermal Expansion of Different Configurations of Fiber-Reinforced Epoxy Composites with $V_f = 60\%$ (Freeman and Kuebeler 1974)

Material Configuration	Coefficient of Thermal Expansion (10^{-6} m/m/°C)		Quasi-Isotropic
	Longitudinal	Transverse	
S-glass–epoxy	6.3	19.8	10.8
Kevlar 49–epoxy	−3.6	54	−0.9 to 0.9
High-modulus carbon epoxy	−0.9	27	0–0.9
Ultrahigh-modulus carbon epoxy	−1.44	30.6	−0.9 to 0.9
Boron–epoxy	4.5	14.4	3.6–5.4
Epoxy	−	54–90	−

3.7.8 THERMAL CONDUCTIVITY

The thermal conductivity of a substance is a measure of how well it can transfer heat to another object. Insulators are a common use for polymers since the group as a whole has low thermal conductivities. However, there are circumstances in which they can also function as heat sinks. This means that they do not dissipate heat as effectively as other materials, which leads to an increase in temperature within the material itself.

The thermal conductivity of a fiber-reinforced polymer is affected by a number of parameters, some of which include the kind of fibers used, their orientation with respect to one another, the volume percentage of the fibers, and the arrangement of the layers. Table 3.7 includes some example values, and it demonstrates that, in general, fiber-reinforced polymers have low thermal conductivities, with the exception of carbon fiber-reinforced polymers, which have high thermal conductivities owing to the highly conductive nature of carbon fibers. Table 3.7 also indicates that carbon fiber-reinforced polymers have low thermal conductivities.

TABLE 3.7
Thermal Conductivity of Different Configurations of Fiber-Reinforced Epoxy Composites with $V_f = 60\%$ (Freeman and Kuebeler 1974)

Material Configuration	Coefficient of Thermal Expansion (W/m per °C)		Quasi-Isotropic
	Longitudinal	Transverse	
S-glass–epoxy	6.3	19.8	10.8
Kevlar 49–epoxy	−3.6	54	−0.9 to 0.9
High-modulus carbon epoxy	−0.9	27	0 to 0.9
Ultrahigh-modulus carbon epoxy	−1.44	30.6	−0.9 to 0.9
Boron–epoxy	4.5	14.4	3.6–5.4
Epoxy	−	54–90	−

When it comes to composites with unidirectional fibers, the thermal conductivity in the longitudinal direction is regulated by the fibers, while the thermal conductivity in the transverse direction is controlled by the matrix. Because of this, the values of the thermal conductivities in these two directions end up being quite different from one another.

The nature of the relationship between the thermal conductivity and the electrical conductivity of fiber-reinforced polymers is quite similar. E-glass fiber-reinforced plastics, for instance, are not good conductors of electricity and have a propensity for accumulating static electricity. Glass fiber composites sometimes include tiny amounts of conductive fibers like carbon fibers, aluminum flakes, or aluminum-coated glass fibers added to them. This helps minimize static charge accumulation, which may then lead to electromagnetic interference or radio frequency interference.

3.7.9 CROSSLINK DENSITY IN SHAPE MEMORY POLYMERS

Swelling or extraction experiments can be used to determine the degree of a crosslinking reaction. This is indicated by the gel content (w_G), which is calculated by dividing the mass of the dried, extracted sample (m_d) by the mass of the original, unextracted sample (m_{iso}), given by Eq. 3.19:

$$w_G = \frac{m_d}{m_{iso}} \times 100\% \tag{3.19}$$

Crosslinking is considered good if values above 90% are achieved. The degree of crosslinking can be determined by measuring the volumetric degree of swelling (Q) in swelling experiments, as stated by Eq. 3.20 as:

$$Q = 1 + \rho_2 \left(\frac{m_{sw}}{m_d \times \rho_1} - \frac{1}{\rho_1} \right) \tag{3.20}$$

Swollen state and dry state sample masses were denoted by m_{sw} and m_d, respectively. ρ_1, ρ_2 are the swelling samples and polymers specific densities.

This is calculated using the mass of the model in its swollen (m_{sw}) and dry (m_d) extracted states, as well as the specific densities of the swelling medium (ρ_1) and the polymer measured using a pycnometer (ρ_2). The Flory–Rehner equation can be used to calculate the average molecular mass of the network chains between two adjacent network nodes ($\bar{M}_c$) and the crosslink density (v_c) of crosslinked polymer networks based on the swelling measurements. For tetra-functional networks, the crosslink density (v_c) can be calculated using Eq. 3.21, where ϕ_2 is the swollen simple volume portion and solvent molar volume denoted by V_1.

$$v_c = \frac{\ln(1 - \phi_2) + \phi_2 + \phi_2^2 \chi_{12}}{V_1 \left[\left(\phi_2/2 - \phi_2^{1/3} \right) \right]} = \frac{\rho}{M_c} \tag{3.21}$$

The Flory solvent-polymer interaction parameter (χ_{12}) and the network density (ρ) are involved in the equation to determine the volume portion of the matrix in the swollen state. This can be done experimentally by calculating the volume of the swollen sample (V_{sw}) and of the dry sample (V_d), with the volume fraction (ϕ_2) being calculated as $\phi_2 = V_d/V_{sw}$.

The crosslink density (v_c) and average mass of molecule of network chains between nearby network nodes ($\bar{M}_c$) for crosslinked polymer networks can also be determined from stress–strain diagrams, based on theories for the rubber elasticity of polymeric networks. When the polymer chains that make up an elastomer are in a state of relaxation, they will form strands in a random pattern. During the process of extension, the chains undergo a stretching action that lowers their conformational entropy. When the stress is removed from the system, the decreased entropy causes the long polymer chains to "regain" their original positions (this phenomenon is known as "entropy elasticity"). The stress (σ) can be experimentally measured, and classical statistical representations of entropy elasticity, such as the affine or phantom network model, derive a simple Eq. 3.22 for σ.

$$\sigma = G\left(\lambda - \frac{1}{\lambda^2}\right) \tag{3.22}$$

In the given equation, λ represents the extension or leeway ratio of the sample, which is equivalent to the ratio of its length L to its original length L_0. It is important to note that the corresponding strain, ε, can be calculated using the formula $\varepsilon = (L - L_0)/L_0$, which simplifies to $\varepsilon = \lambda - 1$. The proportionality constant, G, represents the shear modulus of the model. Eq. 3.22 is particularly effective in describing the behavior of polymer networks subjected to small deformations. Whenever plotting the low elongation stress–strain data, only check the crosslinked property evaluation (Eq. 3.23). Classical theories suggest that the shear modulus G is directly proportional to both the temperature and the crosslink density v_c. This is the case because the shear modulus G is directly proportional to both of these factors.

$$\sigma = v_c RT = \frac{\rho RT}{\bar{M}_c} \tag{3.23}$$

The Mooney–Rivlin equation provides an alternative model for explaining stress–strain data in cases of larger deformations. The rubber elasticity of a polymer structure may be characterized using this equation if it is assumed that the elastomeric sample is both inflexible and isotropic in its natural condition, and that it behaves like a Hookean solid when subjected to simple shear. The behavior of the sample may be analyzed by graphing the observed stress (σ) divided by a factor determined from classical models against the reciprocal deformation ($1/\lambda$) in a Mooney–Rivlin plot of a uniaxial deformation, as stated in Eq. 3.24. This will create a Mooney–Rivlin plot of a uniaxial deformation.

$$\frac{\sigma}{\lambda - \frac{1}{\lambda^2}} = 2C_1 + \frac{2C_2}{\lambda} \tag{3.24}$$

Mooney–Rivlin plot (with $C_2 = 0$) shows that the plot will be a straight horizontal line according to the classical elastic rubbery approach prediction. However, experimental statistics show a positive slope ($C_2 > 0$), indicating stress softening with increasing distortion. By comparing Eqs. 3.22 and 3.24, it becomes evident that for classical models, the Mooney–Rivlin coefficient $2C_1$ corresponds to the shear modulus G, as given by Eq. 3.23. Additionally, the Mooney–Rivlin plot provides an alternative approach to determining crosslink density v_c or the average molecular mass of a linkage chain $\bar{M}_c$.

3.7.10 NUCLEAR MAGNETIC RESONANCE SPECTROSCOPY

NMR spectroscopy can be used to look into the number of reiterating units and their dispersion, molecular mass or weight, branching, tactility, and nonreacted monomers. Both the solid and expanded states of polymer networks are suitable for NMR analysis. The identification of specified baseline-separated signals serves as the foundation for both kinds. NMR spectroscopy of polymer materials in their expanded state only offers information on the macromolecules, while solid-state NMR may also provide information on the state of programming of shape memory polymers (SMPs; Webb and Aliew 2006; Bertmer et al. 2005).

It is possible to make the polymer networks (molar concentration of components) when samples are mined in a fluid for purification by high-resolution magic-angle spinning (HRMAS)-NMR spectroscopy. On AB polymer links with polycaprolactone (PCL) and poly(cyclohexyl methacrylate) pieces, shape memory effect (SME) was displayed (Behl et al. 2009). The weight composition of PCL could be calculated using baseline-separated signals from ^{1}H HRMAS-NMR spectra that were integrated independently to derive concentrations for protons from the cyclohexyl ring's methane group and the methylene group next to the oxygen atom. It was possible to show that the PCL content of the extracted AB polymer networks was nearly equal to the PCL dimethacrylates ingredients of the product combination.

3.7.11 THERMAL CHARACTERIZATION

One of the most important features of SMPs is their temperature performance, particularly T_{trans} connected to the switching domains. Numerous SMPs are phase-segregated polymers, i.e., they have hard and movable domains with distinct T_{trans} associated with each domain. In this part, two techniques for figuring out the temperature characteristics of SMPs are presented: dynamic mechanical thermal analysis (DMTA) and DSC.

3.7.11.1 Dynamic Mechanical Thermal Analysis

This method is used to describe the dynamic properties of polymers. In most cases, a load cell detects force, and an actuator applies an oscillatory displacement. When the sample is vibrated in different directions, such as tension, compression, shear, or twisting, the frequency of the oscillating mechanical stimuli may be altered. The parameters that are determined as a consequence are referred to as the storage modulus E′, the loss modulus E″, and the loss factor $\tan\delta = E''/E'$. The thermal transitions of the polymer may be found by carrying out a DMTA experiment across a range of temperatures at a constant rate of heating or cooling. This will allow for the discovery of transitions. A polymer network experiences a glass transition when it transitions from a rubbery (low modulus) state at low temperatures to a crystalline state (high modulus) at higher temperatures (Huskić et al. 2022). Figure 3.25 depicts the variation of storage modulus changes as a function of temperature for four distinct kinds of SMPs. Measurements of the DMTA were carried out at 1 Hz with a minuscule oscillatory displacement. The two thermoplastic SMPs shown in Figure 3.25 both have crystalline, rigid regions that are heated to a T_m. It is always advisable to combine the determination of steps in the distinctive $E'(T)$-plot as coming from a T_g or T_m with the corresponding DSC tests. Glass transitions have a high frequency dependence, whereas melting is typically frequency independent.

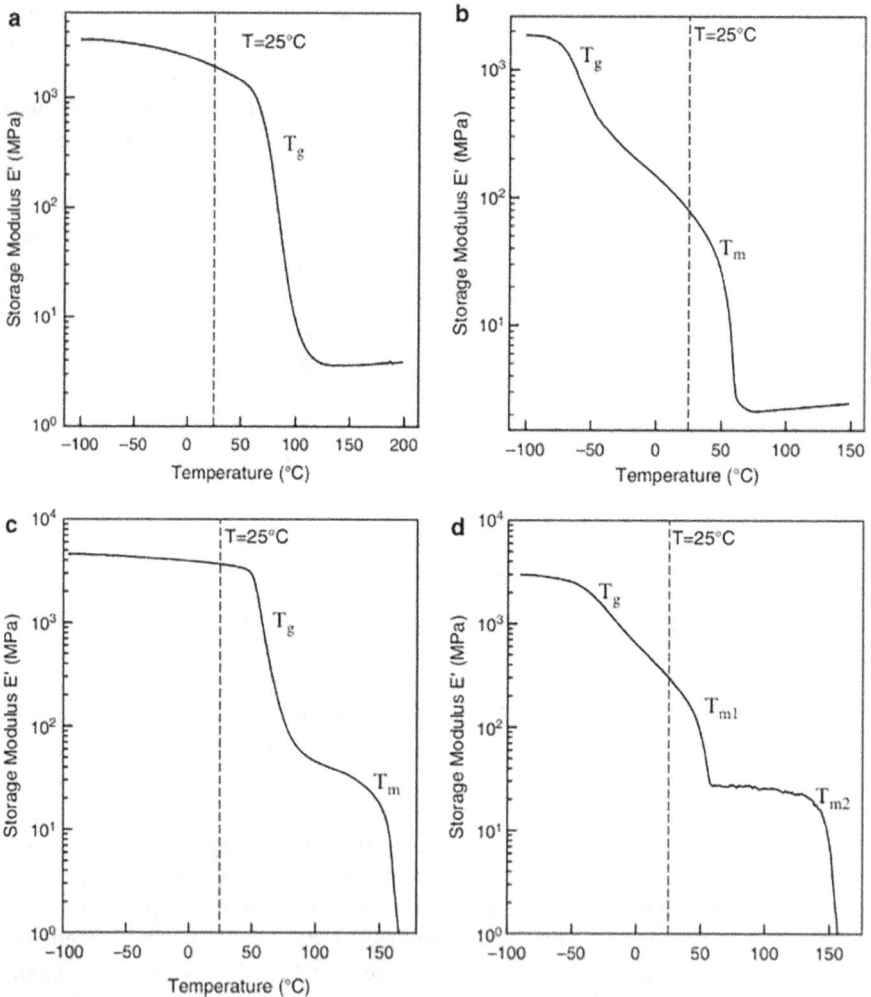

FIGURE 3.25 Shape memory effect of different SMPs with respect to dynamic thermomechanical behavior (Liu, Qin, and Mather 2007).

3.7.11.2 Differential Scanning Calorimetry

The differential scanning calorimeter, or DSC, is a piece of thermal analysis equipment that measures how the physical characteristics of a sample change in conjunction with temperature during the course of an experiment. In other words, the device is an instrument for thermal analysis, and its purpose is to calculate the temperature and heat flow associated with material transitions based on the temperature and the amount of time that has passed. A method in which the temperature of the sample unit, which is constituted by a sample and a reference material, is modified in accordance with a predetermined program, and the temperature difference that exists between the sample and the reference material is measured in accordance with its temperature (Figure 3.26). The molar Gibbs energy's first derivatives are irregular and are studied using DSC (Mark et al. 2004).

Changes in heat conductivity may be used to determine whether a material has gone from a glass state to a viscoelastic state using a technique called DSC. This transition is a second-order endothermic process, and it performs as a step in the DSC plot. In contrast, first-order transitions like melting exhibit a peak in the DSC plot. The modulation point of the stage transition in DSC indicates the glass transition temperature (T_g), but DMTA measurements may provide a better identification of T_g. In SMPs with a liquid crystalline changeover, due to long-range changes in the order parameter, there is a large rise in the specific heat capacity that ultimately leads to the transition temperature. At the transition temperature, a first-order phase transition occurs, resulting in a DSC peak that is a mixture of these two contributions. For example, a liquid crystalline polymer may exhibit sharp endothermic peaks at T_{c-n} for the crystal-nematic transition and T_{n-i} for the nematic–isotropic transition, as shown in Figure 3.27e.

The temperature history of polymer samples as they are processed may have an effect on their characteristics, including their crystallinity and thermomechanical properties. In DSC investigations, the second heating run is usually analyzed after the thermal history of the sample has been erased by heating the sample to a temperature that is higher than the maximum T_{trans} reached in the first run. Analyzing the measured latent fusion heat ΔH_m, which is the area under the curve of the melting peak above the baseline, may be used to assess the crystallinity of a polymer sample. This can be done by comparing the sample to a baseline. To determine the

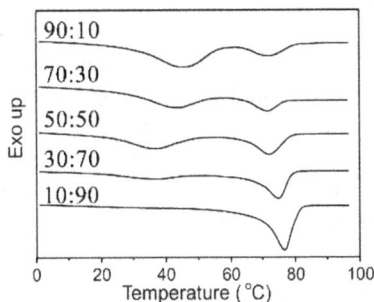

FIGURE 3.26 Illustrations of the thermal behavior of different types of SMPs as characterized by differential scanning calorimetry (Zhang et al. 2023).

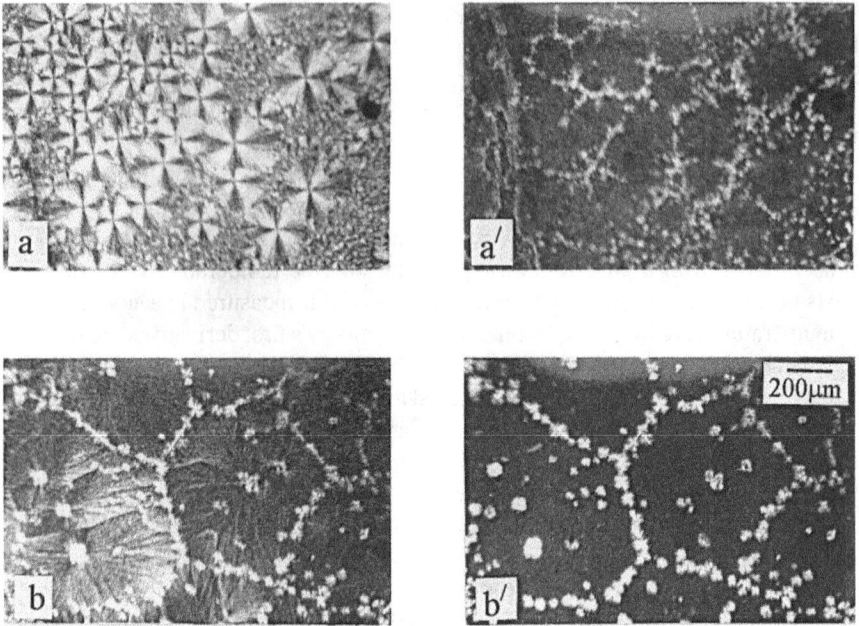

FIGURE 3.27 Polarizing light micrographs for EOET6000–20 (a, a′) and EOET6000–30 (b, b′) samples, well-crystallized at 120 and 35°C for 7 and 15 hours, respectively (Wang, Zhang, and Ma 1999).

crystallinity of the material, $X_c = \Delta H_m / \Delta H_m^0$ is first calculated using this number, and then it is divided by the heat of fusion for a material that is 100% crystalline.

A method known as modulated DSC uses the same equipment as traditional DSC, but instead of applying the same heating and cooling profile to the sample and reference materials, it creates a unique heating and cooling profile for each. In this method, a sinusoidal wave is superimposed on top of the more conventional linear temperature ramp. When this is done, it is possible to manipulate three experimental factors in order to improve the accuracy of DSC measurements. These variables are the frequency, the modulation amplitude, and the heating rate. When compared to conventional DSC, modulated DSC has the advantage of being able to directly measure heat capacity in addition to providing information on both the reversing and nonreversing features of thermal events. This information is not provided by conventional DSC. This method lessens the impact of the baseline slope and curvature, which ultimately results in an increased degree of sensitivity across the system. Molecular relaxation and glass transitions are two examples of overlapping effects that may be distinguished using modified DSC.

3.7.12 MORPHOLOGICAL CHARACTERIZATION

In order to achieve the SME in physically crosslinked SMPs, it is crucial to have the correct morphology. To investigate the morphology of SMPs at different hierarchical levels, various microscopy methods can be employed, such as: (i) polarized light microscopy (POM), (ii) atomic force microscopy, (iii) transmission electron microscopy (TEM), and (iv) scanning electron microscopy (SEM). The selection of the microscopy method depends on the specific morphology of the polymer system and its essential parameters. POM allows for micrometer-scale (spherulite) analysis of SMPs, including crystalline domains. Phase-segregated morphology in SMPs often has to be defined in terms of domain sizes in the nanometer- and micrometer-range, which may be performed by SEM or TEM. Additionally, nanoscale SMP morphology may be depicted by AFM. To accurately define and comprehend the SME of various SMP systems, a complete characterization of the structure and size of domains, as well as information on the kind of crosslinks and crystallizing nature of change, is required.

3.7.12.1   Polarized Light Microscopy with Light

When observed under a microscope using polarized light, many plastics exhibit birefringence. POM utilizes this phenomenon, resulting in pictures that display a distinct brightness for areas with varying degrees of molecule order. Structural integrity into the semicrystalline matrix and hence SMPs studied by POM (Gedde 1995). POM was used to examine the shape of spherulite in ethylene oxide and ethylene terephthalate (EOET) copolymers (Luo et al. 1997). In addition to crystallizing on a nanoscale scale, hard domains of EOET could also develop spherulites at the next step up in hierarchy, as depicted in Figure 3.27. Conclusions about the influence of hard-domain material and its level of crystallinity on the SME were drawn from this research. Higher hard-domain contents were overall better aggregated by creating physical crosslinks, and the associated distortion retrieval was greater, if they had the same swapping segment span but were of a higher hard-domain content. Thermal changes in a polydomain nematic network with a glass-forming texture were studied by POM and determined (Qin and Mather 2009). Experiments with hot-stage POM allowed the detection of two nematic–isotropic transitions.

3.7.12.2   Atomic Force Microscopy

By quantifying the attractive or repulsive forces that exist between a sensor and the model, atomic force microscopy accurately captures the surface texture of materials (Gedde 1995). A raster scan is used to move a fine tip over the sample, and vertical deflections brought on by surface differences are observed. These surface differences in SMPs are dependent on areas with various molecule structures and morphologies.

Zhang et al. studied a blend of two polymers, SBS (styrene–butadiene–styrene) and PCL, and how their phases behave as the content of PCL increases in the blend (H. Zhang et al. 2009) (Figure 3.28). SBS and PCL are immiscible polymers, resulting in phase separations between the two components in the blend. This results in the creation of a blend that will have distinct regions or phases of each polymer. As the content of PCL increases in the blend, its phase behavior changes. At low PCL content (below 20 wt%), PCL exists as a dispersed phase in the form of droplets, surrounded by the continuous phase of SBS, which acts as the matrix. As the PCL content increases, the droplet-like PCL phase becomes more continuous, eventually forming a continuous phase of its own. At even higher PCL content, the PCL phase becomes the matrix, with SBS as the dispersed phase.

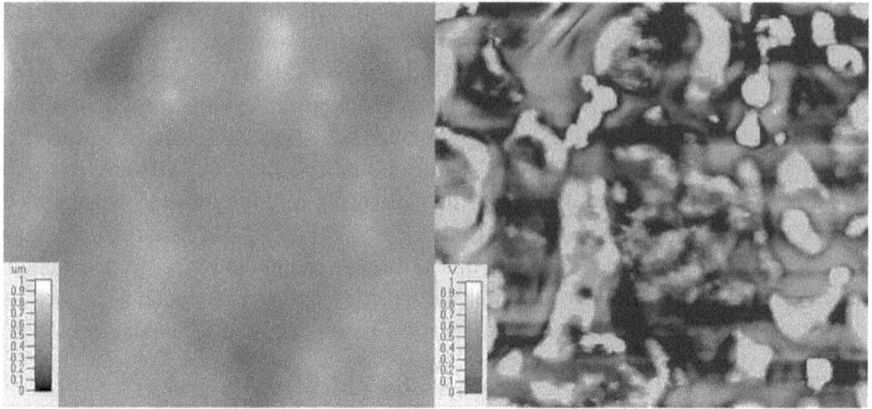

FIGURE 3.28 AFM images of the cross sections of typical SBS/PCL blend samples. (a) PCL20 and (b) PCL30 (Zhang et al. 2009).

3.7.12.3 Scanning and Transmission Electron Microscopy

A concentrated stream of electrons irradiates the sample in an electron microscope and produces a greatly enlarged picture (Cowie and Arrighi 2007). In an SEM, the beam moves across the specimen's surface as the right detectors gather secondary and backscattered electrons. The measurement of polymer surfaces and the establishment of surface contours are accomplished using SEM. For the electrons to enter the material in TEM without losing much energy, it must be very narrow (less than 100 nm). A picture is contrasted when the electron intensity varies in areas with various morphologies. By staining polymers with a heavy metal that adheres primarily to one of the phases present in the sample, the contrast can be improved.

The utilization of SEM was employed to analyze the surfaces resulting from cryogenic fragmentation of graft copolymers comprising nylon 6 and polyethylene. The heterogeneous chemical composition of the copolymers indicated that their creation could affect their overall morphology, resulting in the phase separation of their constituents (Li et al. 1998). In a polyethylene matrix, the nylon fragments exhibited a spherical shape with sizes ranging on the micrometer scale. The primary conclusion drawn from this research was that graft copolymers, as well as linear multiblock copolymers, have the potential to produce thermoplastic SMPs, which was confirmed through SEM. It was crucial in gaining insight into the shape of SMP composites, as demonstrated in the examination of nanocellulose-reinforced shape memory polyurethanes, where it was possible to evaluate the quality of nanofibril distribution. The findings facilitated the calculation of the nanofibril's effect on the SME.

The micro-morphology of a sample was detected through TEM in RuO_4-stained shape memory polyurethanes (SMPUs) (Lin and Chen 1998). 4,4'-methylene diphenyl diisocyanate (MDI), 1,4-butanediol (BD), and poly(tetramethyl oxide)glycol (PTMO) were used to produce SMPUs, with PTMO serving as the lenient domain and MDI/BD as the stiff domain. Various mole ratios of MDI and BD were utilized to investigate the impact on the hard-domain material. The TEM micrograph of the dyed sample with an unaltered hard domain did not exhibit a contrasted image. Conversely, a sample containing both hard and soft domains was subjected to a DSC analysis, which revealed the existence of two distinct domains, with the hard domain being crystallized. Consequently, the TEM image of this material exhibited a contrast pattern. RuO_4 generally causes discoloration in polymer systems, including the MDI/BD hard domain (Trent, Scheinbeim, and Couchman 1983), which contains aromatic compounds as part of its unit structure. The continuous phase in the RuO_4-stained SMPUs research had a network structure and was the hard domain, whereas the dispersed phase, which was abundant in soft segments but remained unstained, was the soft domain.

3.7.13   SCATTERING TECHNIQUES

In order to obtain quantitative structure data for polymers linked to various length scales from the nanometer to the micrometer level, scattering techniques using X-rays or neutrons are used (Stribeck 2007). If crystalline regions are involved in the SME, X-ray scattering is particularly well adapted to study SMPs. A quantified depiction of crystalline hard and switching regions in terms of crystal structure, size, and orientation is possible using wide- and small-angle X-ray scattering (WAXS and SAXS).

3.7.13.1   Wide-Angle X-ray Scattering

WAXS patterns can provide valuable information about the microstructure of polymer chains, including details on crystalline structure, crystal size, and crystallinity. In a study on polycyclooctene (PCO), WAXS analysis revealed a combination of four crystalline diffraction bands and an amorphous surround in the microstructure, with a nearly constant d-spacing across different levels of peroxide crosslinking (Figure 3.29). However, increasing crosslinking concentration resulted in a decrease in crystallinity due to its restricting impact on crystal development. This effect was more prominent in the triclinic crystal structure than the monoclinic crystal structure, indicating different levels of resistance to crosslinking restriction.

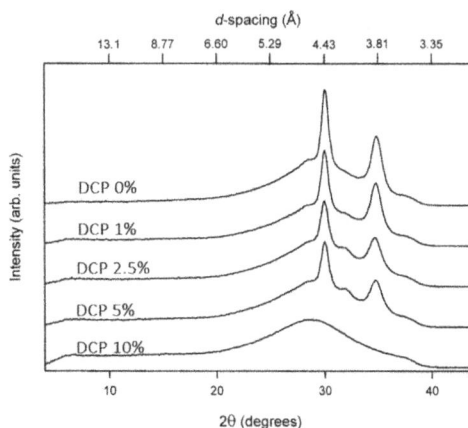

FIGURE 3.29   WAXS profiles of PCO samples with different amounts of peroxide crosslinking: Intensity vs. scattering angle (Changdeng Liu et al. 2002).

WAXS analysis can also be used to understand how crystal structure and formation affect shape-fixing and recovering behavior in semicrystalline SMPs. Additionally, WAXS experiments on polymer fibers from shape memory polyurethane (SMPU) with SME demonstrated that the spinning process can trigger the creation of hard-domain crystallites and lead to a greater degree of crystallinity in shape memory fiber compared to the matching SMPU.

3.7.13.2 Small-Angle X-ray Scattering

One can observe typical nanostructures such as thermoplastic elastomers and lengthy domain sizes in semicrystalline materials through the SAXS regime. On the other hand, USAXS allows the visualization of larger structures like spherulites in the micrometer range. SAXS can be utilized to investigate the organization of crystalline domains within SMPs, along with domain size and orientation as identified by WAXS. The programming and recovery procedures executed on SMPs, such as crystal formation in one phase, significantly affect the orientation and spacing of periodic repeat elements with comparable electron densities within the sample's crystalline domains. To determine the impact of these factors on SME, SAXS studies serve as a valuable technique for characterization.

A study conducted on polyurethane fibers has shown that polyurethanes with a hard segment concentration of approximately 50% form a lamellar morphology (Ji et al. 2006). The long periods of three different types of SMPUs (SMPU-1, SMPU-2, and SMPU-3) were determined to be 10.73, 10.36, and 9.89 nm, respectively, based on the peaks within the curves in Figure 3.30 using Bragg's law. The Lorentz adjusted SAXS intensity profiles of these SMPUs are also presented here. In contrast, the measurements for three types of SMFs (SMF-1, SMF-2, and SMF-3) showed long periods of 10.61, 9.95, and 9.16 nm, respectively, indicating that the SMFs had slightly shorter long periods than the SMPUs. The longer period of the SMFs suggests an increase in the number of hard segment microdomains compared to the aggregate SMPUs.

A study used SAXS to investigate SMPUs containing reduced-molecular-weight switching segments. When examining linear SMPUs based on poly(ethylene adipate) (PEA) using SAXS patterns, a faint scattering maximum was observed, enabling the measurement of long periods for various hard-domain contents (Takahashi, Hayashi, and Hayashi 1996). The PEA segments present in the unclear

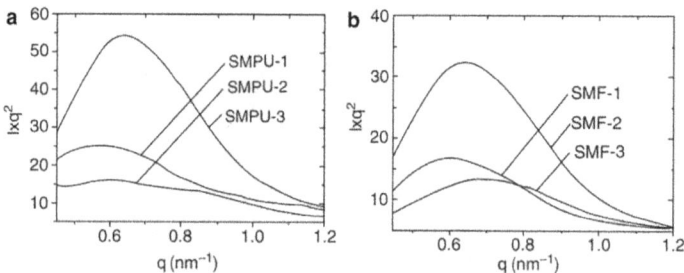

FIGURE 3.30 Small-angle X-ray scattering (SAXS) intensity profiles of polyurethane bulk material (SMPU) and spun fibers (SMFs): Lorentz corrected the SAXS profiles of the SMPUs (a) and the SMFs (b) (Ji et al. 2006).

space between the layered lamellar crystals of the hard domain extended the long period of the SMPUs. It was found that amorphous PEA segments were present in both the amorphous matrix and the amorphous area between the lamellar crystals.

3.7.14 CYCLIC/THERMOMECHANICAL TESTING

Various measurement methods have been discussed in the literature for quantifying an SME, with cyclic thermomechanical tension tests being one of the most effective and popular methods. The SME is quantified using thermomechanical cycle tensile experiments following specific test protocols. These test methods include a programming module for creating the temporary form and a recovery module for recovering the permanent shape. The recovery module can be completed either under continuous pressure or under stress-controlled circumstances, while the programming module can be completed either way (Andreas Lendlein and Kelch 2002). The applied thermomechanical testing parameters, such as applied strain (ε_m), strain rate ($\dot{\varepsilon}$), cooling (β_c) and heating rates (β_h), as well as the applied temperatures for deformation (T_{deform}), fixation of the temporary shape (T_{low}), and recovery of the original permanent shape (T_{high}), all have an impact on the shape memory properties. The programming method can be carried out using various test procedures, including cold drawing and a heating-stretching-cooling process (Khonakdar et al. 2007), with cooling carried out while controlling stress or strain (Choi and Lendlein 2007). Recovery circumstances have been described using different terms such as stress-free, free strain, zero stress, unrestricted, etc., as well as continuous strain, fixed strain, limited strain, stress-generating, etc. (Liu et al. 2006; Yakacki et al. 2007). In this text, the terms "stress-free" and "continuous tension" will be used. While strain-controlled tests detect the change in strain, stress-controlled tests enable the measurement of the specimen's stress at specific thermal conditions. Two common test procedures are described as examples (Figure 3.31).

The recovery curves for thermoplastics and polymeric networks under stress-free and continuous strain conditions are depicted in schematic form in Figure 3.32. For a thermoplastic SMP, the recovery under continuous tension results in an increase in

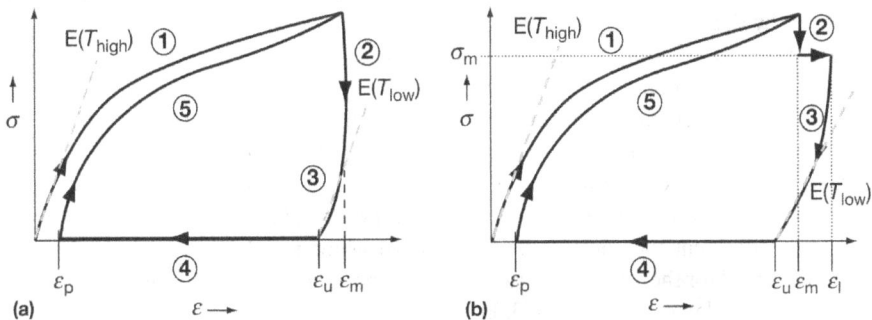

FIGURE 3.31 Schematic representation of the results of the cyclic, thermomechanical tensile tests for two different cycle types. (a) Strain-controlled programming with stress-free recovery (ε–σ diagram). (b) Stress-controlled programming with stress-free recovery (Lendlein et al. 2017).

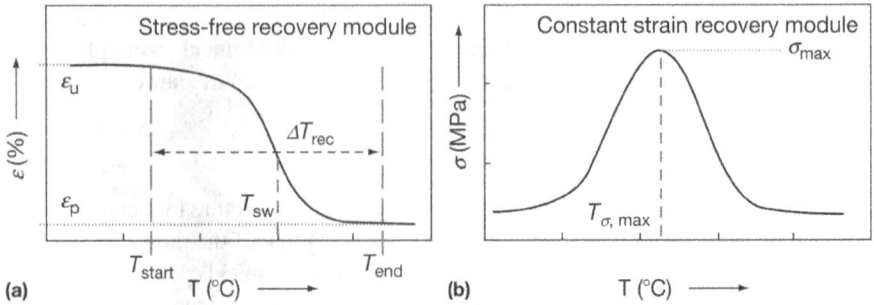

FIGURE 3.32 Schematic representation of recovery curves for thermoplastic SMPs. (a) ε–T diagram of a thermoplastic SMP recovered under stress-free conditions. (b) σ–T diagram of a thermoplastic SMP, recovered under constant strain conditions (Lendlein et al. 2017).

tension until $T_{\sigma\,max}$ is reached. At higher temperatures, the tension decreases due to the relaxation of the matrix, which is initiated by an increase in chain segment mobility. On the other hand, stress maintains a constant zone in the SMP above $T_{\sigma\,max}$ and does not decrease until the sample is cooled down again.

3.8 CASE STUDY: FABRICATION AND CHARACTERIZATION OF MULTIPHASE SHAPE MEMORY COMPOSITES

3.8.1 SPECIMEN PREPARATION

Fabrication of functionalized GNP–epoxy nanocomposite: The researchers incorporated commercially procured graphene nanoplatelets (GNPs) into an epoxy matrix to create GNP–epoxy nanocomposites. To achieve this, the GNP was first dissolved in acetone through magnetic stirring for 30 minutes. The epoxy resin was then added to the mixture and ultrasonicated for 2 hours using an ultrasonic probe. After removing the acetone, the resulting mixture uniformly dispersed GNP within the SMP matrix. The researchers prepared SMP nanocomposites with varying fGNP contents (0.4, 0.6, and 0.8 wt%). Any air bubbles that formed during the mixing process were removed by degassing the mixture in a vacuum chamber.

    Fabrication of shape memory hybrid composites: The researchers added hardener to the GNP mixture using the layup technique to create hybrid laminated composites. In this experiment, they used five laminae with a constant carbon fiber volume fraction of 23%. After stacking the composites in a unidirectional manner, samples were cured in an autoclave at five bar pressure and 80°C. Once the laminated composites reached adequate curing with dimensions of 250 mm × 250 mm, the researchers cut the samples using abrasive water jet machining, following the standards used for the various characterizations performed in this study. The fabrication process is depicted in Figure 3.33.

3.8.2 MATERIAL CHARACTERIZATION

The mechanical characterization of the SMP composites and Shape memory hybrid composite (SMHC) samples was performed with the universal testing machine

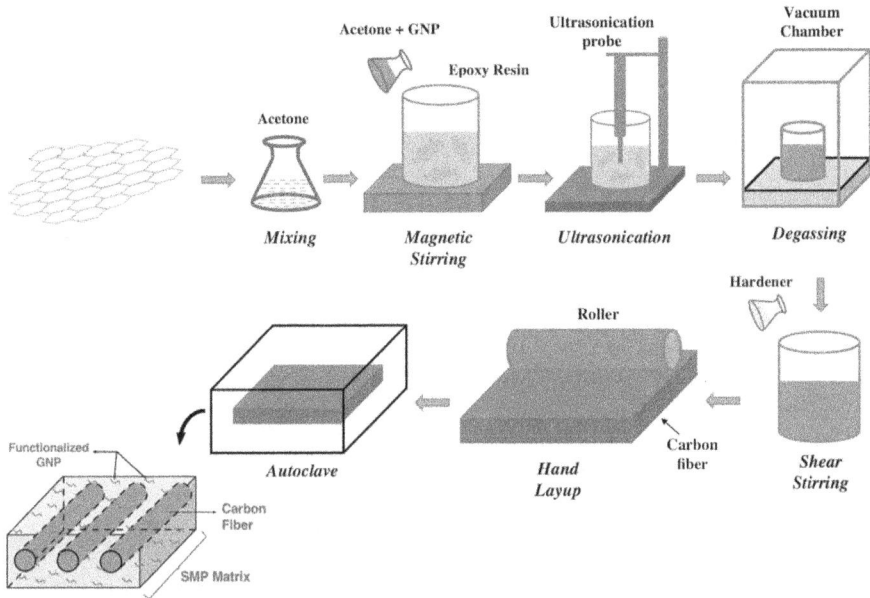

FIGURE 3.33 Schematic layout of the fabrication of shape memory hybrid composites (Tiwari, Shaikh, and Malek 2022).

(UTM) as per ASTM D3039 on UTM5582, Instron (ARAI, Pune). Tensile load at a rate of 5 mm/minute was applied to the composite samples with dimensions of 250 ± 0.1 mm $\times 250 \pm 0.1$ mm $\times 1.5 \pm 0.1$ mm, as per the ASTM standard. The researchers analyzed five specimens of each configuration so that they could assure accuracy and efficiency in the specimen preparation process. In order to do analytical and numerical simulations of the tensile behavior of the material, they examined a variety of mechanical parameters, such as the longitudinal elastic modulus, Poisson's ratio, tensile strength, maximum load, and elongation at failure. These mechanical properties are important for modeling the tensile performance of the material.

The dynamic mechanical analysis (DMA) is used to investigate the prepared samples thermomechanical behavior utilizing 3-point bending in a dynamic mechanical analyzer (DMA 800, Perkin Elmer). The prepared samples had dimensions of 40 ± 0.1 mm $\times 10 \pm 0.1$ mm, and ASTM standards of D 7028-7 were maintained, with the temperature of the samples varying from 30°C to 150°C at a rate of 3°C with a 1 Hz frequency load. For the analysis, results obtained with the test are in the form of storage and loss moduli, tan δ, which are important for suggesting the glass transition temperature (T_g) of the plate and equivalent dynamic bending variational nature of the samples in the glass transition region of the SMP polymer.

The morphological modification caused by the addition of fGNP to the shape memory hybrid composites was studied by analyzing fracture samples obtained from the tensile test using field-emission scanning electron microscopy. In addition to investigating the interfacial bonding between the carbon fiber and the modified matrix, the researchers also assessed the fiber's wettability. The Nova Nano FE-SEM 450-FEI was used to perform the field-emission scanning electron microscopy.

The specimens' shape memory behavior was evaluated using heat-activated bending tests. Each shape memory hybrid composite was sliced into plates measuring 40 ± 0.1 mm $\times$ 10 ± 0.1 mm. It was believed that the initial angle (θ_o) of each rectangular plate was 180°. The specimen was heated to the programming angle (θ_p) of 90° at a temperature of $T_g + 20°$. The shape fixity (R_f) was determined after the deformed sample was slowly cooled, and the original shape was restored by heating the sample to $T_g + 20°$. Recovery (R_r) was measured after taking into account the recovery angle (θ_r). Each sample's form memory cycle was captured with a high-definition camera and then analyzed using CAD software to determine its θ_o, θ_p, and θ_r values. Three replicates of each setup were used to compile this data. Using Eqs. (3.1) and (3.2), we analyzed the data to determine how well each specimen configuration kept its temporary programmed form and how well it returned to its original shape (Nilesh Tiwari and Shaikh 2022b).

$$R_f = \frac{\left|(\theta_f - \theta_0)\right|}{\left|(\theta_p - \theta_0)\right|} \tag{3.25}$$

$$R_r = \frac{\left|(\theta_f - \theta_r)\right|}{\left|(\theta_f - \theta_0)\right|} \tag{3.26}$$

3.8.3 MECHANICAL PROPERTIES

The tensile behavior of carbon fiber-reinforced nanocomposites was examined to determine the influence of varying GNP content on their mechanical properties. Elastic modulus was analyzed for nanocomposites containing 0, 0.4, 0.6, and 0.8 wt% of GNP, using ASTM D 3039 to prepare the samples. Incorporating GNP into the SMP matrix improved the bond between the matrix and carbon fiber. Specifically, modulus increased by 9.7% and 22.6% when 0.4 wt% and 0.6 wt% of GNP were added, respectively, as shown in Figure 3.34. This improvement is likely due to the non-covalent surface bond of uniformly dispersed GNP during ultrasonication and stirring. Additionally, the enhanced dispersion of GNP increased the available surface area, improving interaction with the SMP matrix (Kumar Patel and Purohit 2019).

The superior tensile properties observed in the SMP nanocomposites with 0.4% and 0.6% GNP content may be due to the combined effects of two phenomena: improved material properties of the SMP matrix resulting from GNP incorporation, and enhanced wettability of carbon due to increased interphase adhesion at the fiber–SMP interface. Efficient transfer of stress from the SMP matrix to GNP across the GNP-matrix interphase likely contributed to the enhanced mechanical properties of the SMP matrix (Nilesh Tiwari and Shaikh 2022a). The quality of the GNP-matrix interphase bond is determined by various factors, such as chemical, physical, and mechanical characteristics, which subsequently influence the fiber–matrix interphase. The significant increase in interphase surface area regulates the overall mechanical properties of the carbon fiber-reinforced nanocomposites modified with GNP. Graphene nanoparticles exhibit a remarkable increase in surface

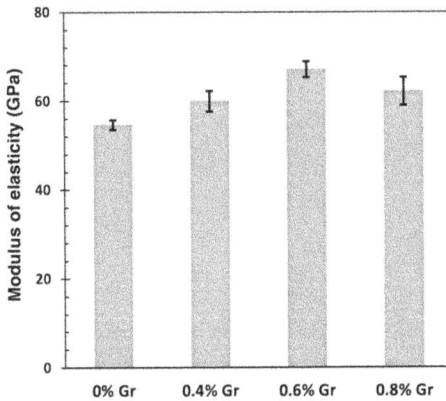

FIGURE 3.34 Tensile testing (ASTM 3039): Effect of variation of %wt of GNP on the modulus of elasticity of nanocomposites (Nilesh Tiwari and Shaikh 2022b).

area compared to unmodified epoxy matrix, ultimately leading to an increased interphase surface area between carbon fiber and the modified matrix. This availability of an extensive interphase area enables efficient transmission of stress from the SMP matrix to GNP and then to the carbon fiber, resulting in the nanocomposite's ability to withstand higher loads and improved tensile properties (Markad and Lal 2021). Nanoparticle-filled polymers possess higher strength than their respective bulk polymers due to the physisorption of the polymer incorporated into the nanoparticle surface. The increase in modulus observed with 0.4% and 0.6% GNP content in the modified nanocomposites indicates the uniform dispersion of GNP in the SMP epoxy matrix, leading to an improved interfacial area between the carbon fiber and modified matrix and a more efficient translation of stress from the SMP matrix to GNP and then to the carbon fiber.

The increase in elastic modulus observed in the SMP nanocomposites with 0.4% and 0.6% GNP content was accompanied by a corresponding improvement in tensile strength of 14.3% and 32.2%, respectively. The enhanced stress transfer from the modified matrix to the carbon fiber due to GNP's presence likely contributed to this improvement. This is also validated by the results in Figures 3.35 and 3.36.

An increase in the percentage of GNP from 0.6% to 0.8% was found to result in a decline in both the modulus and tensile strength. While the modulus of the nanocomposite with 0.8 wt% GNP was still 13.7% higher than that of the unmodified matrix Carbon fiber reinforced polymer (CFRP), it was 7.3% lower than that of the nanocomposite with 0.6 wt% GNP. This degradation in tensile properties could be attributed to the agglomeration of nanoparticles caused by the generation of high surface energy due to increased surface area. The agglomeration leads to a reduction in the available interphase between the modified matrix and the carbon fiber, resulting in a decrement in stress transmission at the interphase, leading to failure at a lower load (Markad and Lal 2022). The morphological studies conducted to analyze the interfacial bonding align with the mechanical behavior results observed in the CFRP nanocomposites with different GNP contents.

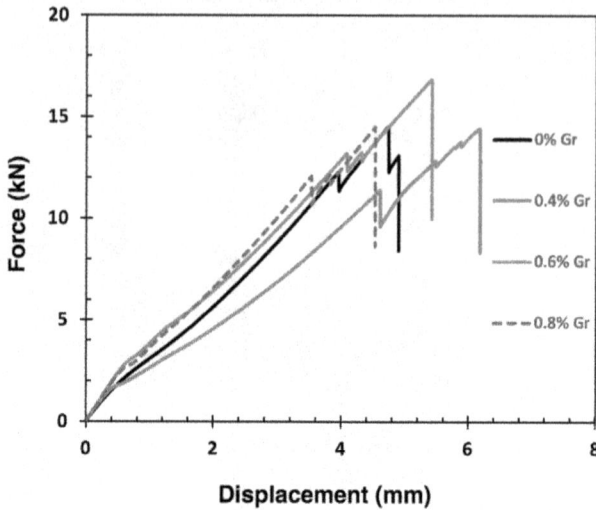

FIGURE 3.35 Effect of variation of GNP on the tensile properties of nanocomposites indicates force with displacement.

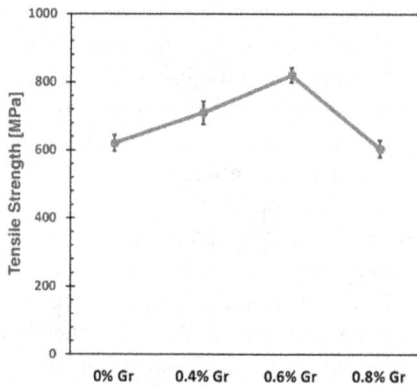

FIGURE 3.36 Mechanical properties with the variation of %wt of GNP tensile strength.

3.8.4 THERMOMECHANICAL ANALYSIS

In this study, the connection between storage modulus (E') with respect to temperature (T) was analyzed using DMA. The temperature range considered in the study was from 30°C to 150°C, and the results are presented in Figure 3.37. The behavior of the storage modulus with respect to the change in GNP content was similar to that observed for the tensile properties earlier in the glass transition region up to 65°C. However, during the glass transition region and after, the variation in behavior was not distinct. The magnitude of storage modulus for different sample configurations

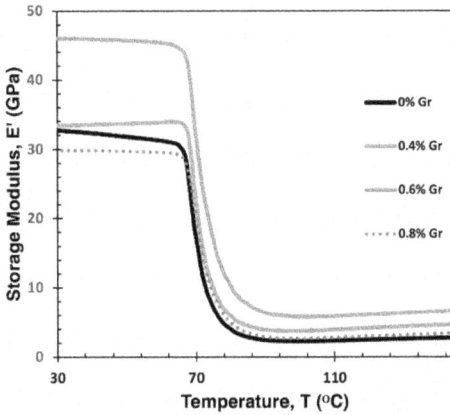

FIGURE 3.37 Thermomechanical analysis of shape memory hybrid composites.

TABLE 3.8
The Assessment of Thermomechanical Properties (Nilesh Tiwari and Shaikh 2022b)

Material Configuration	Storage Modulus before Glass Transition (E′) GPa	Storage Modulus after Glass Transition (E′) GPa	Peak Factor (tan δ)
Carbon fiber + Epoxy	27.1	2.6	0.247
Carbon fiber + Epoxy + GNP	29.9	2.8	0.240

GNP, graphene nanoplatelets.

was similar after the glass transition region, indicating the redundant influence of carbon fiber reinforcement. Table 3.8 provides comparison data for the DMA between unmodified CFRP and hybrid fiber-reinforced polymer.

3.8.4.1 Morphology

To investigate the impact of GNPs on the interaction between the carbon fiber and modified SMP matrix, the surface morphology of failed samples with varying GNP content was examined. The SEM images in Figure 3.38 reveal the effect of GNP variation on interfacial interactions, which are crucial for the mechanical, thermal, and thermomechanical properties of the nanocomposites. Without GNP, the CFRP composites exhibit weak interfacial bonding, resulting in debonding at the interphase. Furthermore, the fibers exhibit a smooth morphology with limited evidence of matrix adherence on the fiber surface. These findings are consistent with the results of mechanical and thermomechanical characterization of the corresponding composites.

The incorporation of GNP in the modified SMP matrix improved the interfacial bond and increased the amount of matrix adhering to the carbon fiber surface. This improvement in adhesion indicates enhanced interactions between the fiber and GNP-modified matrix. In addition, the wettability of the fiber was improved with the addition

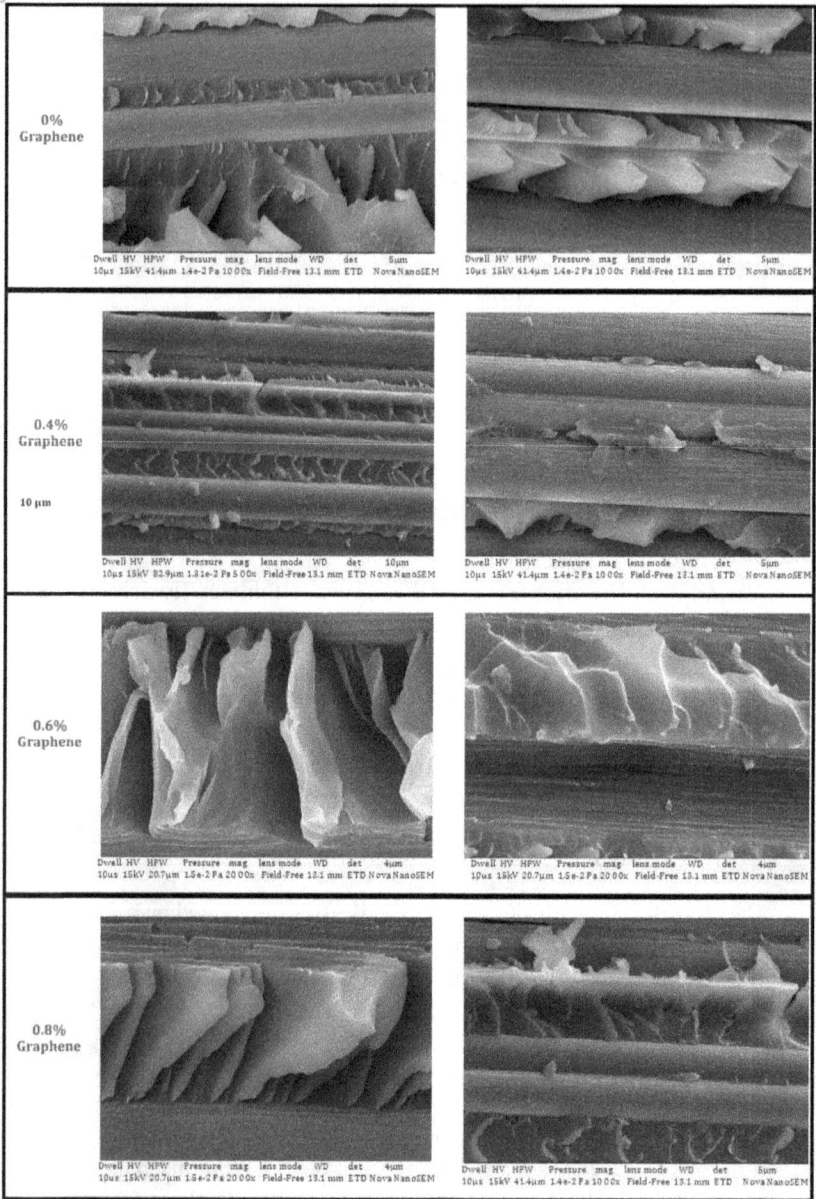

FIGURE 3.38 Morphology of fractured samples with different %wt of GNP (Nilesh Tiwari and Shaikh 2022b).

of nanoparticles to the matrix. The improved adhesive bond between the fiber and matrix resulted in higher resistance to debonding or delamination under mechanical or thermomechanical loadings. This resistance, in turn, led to increased structural integrity of the nanocomposites, thereby enhancing their physical, chemical, and mechanical properties. The surface roughness of the fibers with 0.6% GNP was observed to be greater than that with 0.4%, indicating their superior mechanical and thermomechanical performance. Thus, the improved adhesion and wettability facilitated the efficient transmission of stress from the matrix to GNP and then to the carbon fibers, resulting in a noticeable enhancement in the tensile and thermomechanical characteristics of the nanocomposites.

The SMP nanocomposite containing 0.8% GNP showed uneven adhesive surfaces on the carbon fiber due to scattered traces of modified nanoparticle bundles. This may be attributed to the presence of GNP agglomerates in the matrix, indicating that agglomeration of GNP occurs above 0.5 wt%. Thus, the incorporation of GNP above 0.6 wt% would lead to the formation of agglomerate bundles in the nanocomposite, negatively affecting its mechanical and thermomechanical performance.

Figure 3.39 displays SEM images of the carbon fibers extracted from each configuration of fabricated samples during the tensile failure tests. The fibers from unmodified CFRP composites exhibited a smooth surface, indicative of inferior interfacial bonding compared to the nanocomposite with nanoparticle incorporation. Fibers from nanocomposites with 0.4% and 0.6% GNP incorporation exhibited concave rough surfaces, indicating improved wettability and adhesion between the fiber and matrix. This resulted in enhanced mechanical and thermomechanical properties due to efficient stress transmission at the interphase. However, the samples with 0.8% GNP exhibited similar curvatures scattered throughout, leading to comparatively degraded properties, likely due to the presence of nanoparticle agglomerates.

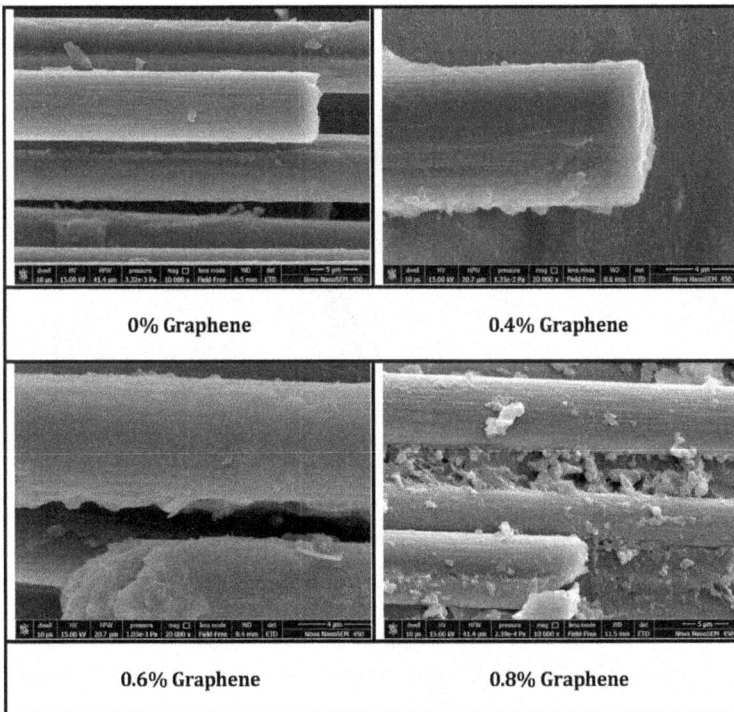

FIGURE 3.39 Field-emission scanning electron microscopy images of the separated carbon fibers from the fractured samples with different %wt of GNP (Nilesh Tiwari and Shaikh 2022b).

3.8.5 SHAPE MEMORY PROPERTIES

The shape memory performance of the hybrid composites was evaluated in terms of shape fixity ratio (R_f) and shape recovery ratio (R_r), and their cycles are presented in Figure 3.40. The heat activation tests were conducted at a temperature of $T_g + 20°C$ for all sample configurations. As the concentration of GNP increased in the modified matrix, a slight delay was observed in the recovery of the samples to their original orientation. This delay may be attributed to the increased resistance offered by the structure due to improved stiffness and interfacial bonding resulting from the incorporation of nanofillers. However, there was no significant effect on the overall values of R_f and R_r.

The unmodified epoxy composites showed complete recovery within 9 seconds, while the hybrid composites experienced an initial delay in recovery but achieved full recovery within 5 seconds. This indicates that the incorporation of GNP in the epoxy matrix improved the thermal response of the SMP composites, reducing the time lag between stress generation and release. Therefore, the addition of 0.4 wt% and 0.6 wt% GNP nanoparticles to the epoxy matrix improved the Young's modulus and storage modulus of the composites without compromising their shape memory properties. However, the composite with 0.8 wt% GNP exhibited delayed recovery similar to the unmodified matrix. This is consistent with the observed decrease in modulus due to nanoparticle agglomeration in the matrix. The values of the shape memory parameters are summarized in Table 3.9.

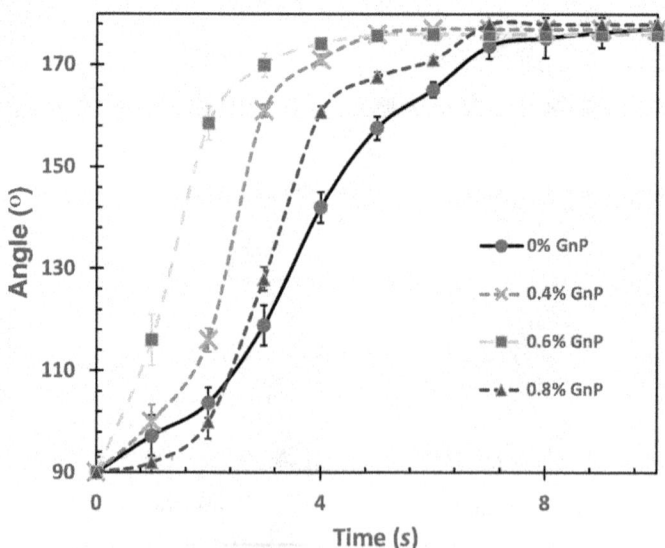

FIGURE 3.40 Bending angle of the samples with respect to time during the recovery stage (Nilesh Tiwari and Shaikh 2022b).

TABLE 3.9
Shape Fixity (R_f) and Shape Recovery (R_r) of SMHC

Material Constituent	R_f(%)	R_r(%)
CFRP + 0% GNP	96.85	98.88
CFRP + 0.4% GNP	97.70	96.66
CFRP + 0.6% GNP	96.59	97.77
CFRP + 0.8% GNP	95.42	97.22

GNP, graphene nanoplatelets.

REFERENCES

Behl, Marc, Ingo Bellin, Steffen Kelch, Wolfgang Wagermaier, and Andreas Lendlein. 2009. "One-Step Process for Creating Triple-Shape Capability of AB Polymer Networks." *Advanced Functional Materials* 19 (1): 102–8. https://doi.org/10.1002/adfm.200800850.

Bertmer, Marko, Alina Buda, Iris Blomenkamp-Höfges, Steffen Kelch, and Andreas Lendlein. 2005. "Solid-State NMR Characterization of Biodegradable Shape-Memory Polymer Networks." *Macromolecular Symposia* 230 (1): 110–5. https://doi.org/10.1002/masy.200551149.

Bhat, Prabhas, Justin Merotte, Pavel Simacek, and Suresh G. Advani. 2009. "Process Analysis of Compression Resin Transfer Molding." *Composites Part A: Applied Science and Manufacturing* 40 (4): 431–41. https://doi.org/10.1016/j.compositesa.2009.01.006.

Chamis, Christos C., and John H. Sinclair. 1978. "Mechanical Behavior and Fracture Characteristics of Off-Axis Fiber Composites. 2: Theory and Comparisons." https://ntrs.nasa.gov/citations/19780008155.

Choi, Nok-young, and Andreas Lendlein. 2007. "Degradable Shape-Memory Polymer Networks from Oligo[(L-Lactide)-Ran-Glycolide]Dimethacrylates." *Soft Matter* 3 (7): 901–9. https://doi.org/10.1039/B702515G.

Cowie, John Mackenzie Grant, and Valeria Arrighi. 2007. *Polymers: Chemistry and Physics of Modern Materials.* CRC press: Boca Raton.

Daniels, Harakas, and Jackson. 1971. "Short beam shear tests of graphite fiber composites." *Fibre Science and Technology* 3 (3): 187–208. https://doi.org/10.1016/0015-0568(71)90002-9.

De Baere, Ives, Wan Van Paepegem, and Joris Degrieck. 2008. "Comparison of the Modified Three-Rail Shear Test and the [(+45°, −45°)]Ns Tensile Test for Pure Shear Fatigue Loading of Carbon Fabric Thermoplastics." *Fatigue & Fracture of Engineering Materials & Structures* 31 (6): 414–27. https://doi.org/10.1111/j.1460-2695.2008.01231.x.

DeTeresa, Steven J., Dennis C. Freeman, and Scott E. Groves. 2004. "The Effects of Through-Thickness Compression on the Interlaminar Shear Response of Laminated Fiber Composites." *Journal of Composite Materials* 38 (8): 681–97.

Dw, Wilson. 1990. "Evaluation of the V-Notched Beam Shear Test Through an Interlaboratory Study." *Composites Technology and Research* 12 (3): 131–8. https://doi.org/10.1520/CTR10189J.

Freeman, William T., and G. C. Kuebeler. *1974. "Composite Materials: Testing and Design (Third Conference)."* ASTM International: Williamsburg.

Gedde, Ulf. 1995. *Polymer Physics.* Springer Science & Business Media: Berlin.

Hahn, Hong T., and Richard Y. Kim. 1976. "Fatigue Behavior of Composite Laminate." *Journal of Composite Materials* 10 (2): 156–80. https://doi.org/10.1177/002199837601000205.

Hahn, Richard Kim. 1975. "Proof testing of composite materials." *Journal of Composite Materials* 9 (3): 297–311.

Huskić, Miroslav, Dragan Kusić, Irena Pulko, and Blaž Nardin. 2022. "Determination of Residual Stresses in Amorphous Thermoplastic Polymers by DMA." *Journal of Applied Polymer Science* 139 (48): e53210. https://doi.org/10.1002/app.53210.

Ji, FengLong, Yong Zhu, JinLian Hu, Yan Liu, Lap-Yan Yeung, and GuangDou Ye. 2006. "Smart Polymer Fibers with Shape Memory Effect." *Smart Materials and Structures* 15 (6): 1547. https://doi.org/10.1088/0964-1726/15/6/006.

Joshi, S. C. 2012. "12- The Pultrusion Process for Polymer Matrix Composites." In *Manufacturing Techniques for Polymer Matrix Composites (PMCs)*, edited by Suresh G. Advani and Kuang-Ting Hsiao, 381–413. Woodhead Publishing Series in Composites Science and Engineering. Woodhead Publishing (U.K.). https://doi.org/10.1533/97808 57096258.3.381.

Khonakdar, Hossein Ali, Seyed Hassan Jafari, Sorour Rasouli, Jalil Morshedian, and Hossein Abedini. 2007. "Investigation and Modeling of Temperature Dependence Recovery Behavior of Shape-Memory Crosslinked Polyethylene." *Macromolecular Theory and Simulations* 16 (1): 43–52. https://doi.org/10.1002/mats.200600041.

Kumar Patel, Krishan, and Rajesh Purohit. 2019. "Improved Shape Memory and Mechanical Properties of Microwave-Induced Thermoplastic Polyurethane/Graphene Nanoplatelets Composites." *Sensors and Actuators A: Physical* 285 (January): 17–24. https://doi.org/10.1016/j.sna.2018.10.049.

Lendlein, Andreas, Muhammad Yasar Razzaq, Christian Wischke, Karl Kratz, Matthias Heuchel, J. Zotzmann, Bernhard Hiebl, A. T. Neffe, and M. Behl. 2017. "Shape-Memory Polymers." *Comprehensive Biomaterials II - Reference Module in Materials Science and Materials Engineering, Metallic, Ceramic, and Polymeric Biomaterials* 1: 620–47. https://doi.org/10.1016/B978-0-12-803581-8.10213-9.

Lendlein, Andreas, and Steffen Kelch. 2002. "Shape-Memory Polymers." *Angewandte Chemie International Edition* 41 (12): 2034–57. https://doi.org/10.1002/1521-3773(20020617)41: 12<2034::AID-ANIE2034>3.0.CO;2-M.

Li, Fengkui, Yan Chen, Wei Zhu, Xian Zhang, and Mao Xu. 1998. "Shape Memory Effect of Polyethylene/Nylon 6 Graft Copolymers." *Polymer* 39 (26): 6929–34. https://doi.org/10.1016/S0032-3861(98)00099-8.

Lin, J. R., and L. W. Chen. 1998. "Study on Shape-Memory Behavior of Polyether-Based Polyurethanes. I. Influence of the Hard-Segment Content." *Journal of Applied Polymer Science* 69 (8): 1563–74. https://doi.org/10.1002/(SICI)1097-4628(19980822)69:8<1563: :AID-APP11>3.0.CO;2-W.

Liu, Changdeng, Haihu Qin, and Patrick T. Mather. 2007. "Review of Progress in Shape-Memory Polymers." *Journal of Materials Chemistry* 17 (16): 1543–58. https://doi.org/10.1039/B615954K.

Liu, Changdeng, Seung B. Chun, Patrick T. Mather, Lei Zheng, Elisabeth H. Haley, and E. Bryan Coughlin. 2002. "Chemically Cross-Linked Polycyclooctene: Synthesis, Characterization, and Shape Memory Behavior." *Macromolecules* 35 (27): 9868–74. https://doi.org/10.1021/ma021141j.

Liu, Yiping, Ken Gall, Martin L. Dunn, Alan R. Greenberg, and Julie Diani. 2006. "Thermomechanics of Shape Memory Polymers: Uniaxial Experiments and Constitutive Modeling." *International Journal of Plasticity* 22 (2): 279–313. https://doi.org/10.1016/j.ijplas.2005.03.004.

Luo, Xiaolie, Xiaoyun Zhang, Mingtai Wang, Dezhu Ma, Mao Xu, and Fengkui Li. 1997. "Thermally Stimulated Shape-Memory Behavior of Ethylene Oxide-Ethylene Terephthalate Segmented Copolymer." *Journal of Applied Polymer Science* 64 (12): 2433–40. https://doi.org/10.1002/(SICI)1097-4628(19970620)64:12<2433::AID-APP17>3.0.CO;2-1.

Makeev, Andrew. 2013. "Interlaminar Shear Fatigue Behavior of Glass/Epoxy and Carbon/ Epoxy Composites." *Composites Science and Technology* 80: 93–100. https://doi. org/10.1016/j.compscitech.2013.03.013.

Mallick, Pankar K. 1981. "Fatigue Characteristics of High Glass Content Sheet Molding Compound (SMC) Materials." *Polymer Composites* 2 (1): 18–21. https://doi.org/10.1002/ pc.750020105.

Mallick, Pankar K. 2007. *Fiber-Reinforced Composites: Materials, Manufacturing, and Design.* CRC press: Boca Raton.

Mark, James, Kia Ngai, William Graessley, Leo Mandelkern, Edward Samulski, George Wignall, and Jack Koenig. 2004. *Physical Properties of Polymers.* Cambridge University Press: England.

Markad, Kanif, and Achchhe Lal. 2021. "Experimental Investigation of Shape Memory Polymer Hybrid Nanocomposites Modified by Carbon Fiber Reinforced Multi-Walled Carbon Nanotube (MWCNT)." *Materials Research Express* 8 (10): 105015. https://doi. org/10.1088/2053-1591/ac2fcc.

Markad, Kanif, and Achchhe Lal. 2022. "Synthesis of the Multiphase Shape Memory Hybrid Composites Hybridized with Functionalized MWCNT to Improve Mechanical and Interfacial Properties." *Polymer-Plastics Technology and Materials* 61 (6): 650–64. https://doi.org/10.1080/25740881.2021.2006709.

Mouritz, Adrian P. 2012. "Manufacturing of Fibre-Polymer Composite Materials." *Introduction to Aerospace Materials* 10: 303–37.

Phillips, D. C., and J. M. Scott. 1977. "The Shear Fatigue of Unidirectional Fibre Composites." *Composites, Fatigue of frp composites,* 8 (4): 233–6. https://doi. org/10.1016/0010-4361(77)90108-2.

Pipes, R. Byron. 1974. *Interlaminar Shear Fatigue Characteristics of Fiber-Reinforced Composite Materials.* ASTM International: Philadelphia, PA.

Qin, Haihu, and Patrick T. Mather. 2009. "Combined One-Way and Two-Way Shape Memory in a Glass-Forming Nematic Network." *Macromolecules* 42 (1): 273–80. https://doi. org/10.1021/ma8022926.

Quanjin, Ma, M. R. M. Rejab, M. S. Idris, B. Bachtiar, J. P. Siregar, and M. N. Harith. 2017. "Design and Optimize of 3-Axis Filament Winding Machine." *IOP Conference Series: Materials Science and Engineering* 257 (1): 012039. https://doi. org/10.1088/1757-899X/257/1/012039.

Rousseau, Judith, Dominique Perreux, and Nathalie Verdière. 1999. "The Influence of Winding Patterns on the Damage Behaviour of Filament-Wound Pipes." *Composites Science and Technology* 59 (9): 1439–49. https://doi.org/10.1016/S0266-3538(98)00184-5.

Shrigandhi, Ganesh D., and Basavaraj S. Kothavale. 2021. "Biodegradable Composites for Filament Winding Process." *Materials Today: Proceedings, 3rd International Conference on Materials Engineering & Science* 42 (January): 2762–8. https://doi.org/10.1016/j. matpr.2020.12.718.

Sofi, Tasdeeq, Stefan Neunkirchen, and Ralf Schledjewski. 2018. "Path Calculation, Technology and Opportunities in Dry Fiber Winding: A Review." *Advanced Manufacturing: Polymer & Composites Science* 4 (3): 57–72. https://doi.org/10.1080/20550340.2018.1500099.

Stribeck, Norbert. 2007. *X-Ray Scattering of Soft Matter.* Springer Science & Business Media: Berlin.

Takahashi, Toshisada, Noriya Hayashi, and Shunichi Hayashi. 1996. "Structure and Properties of Shape-Memory Polyurethane Block Copolymers." *Journal of Applied Polymer Science* 60 (7): 1061–9. https://doi.org/10.1002/(SICI)1097-4628(19960516)60:7<1061::AID-APP18>3.0.CO;2-3.

Tiwari, Nilesh, AbdulHafiz A. Shaikh, and Naved I. Malek. 2022. "Modification of the Multiphase Shape Memory Composites with Functionalized Graphene Nanoplatelets: Enhancement of Thermomechanical and Interfacial Properties." *Materials Today Chemistry* 24: 100826. https://doi.org/10.1016/j.mtchem.2022.100826.

Tiwari, Nilesh, and AbdulHafiz A. Shaikh. 2022a. "Effect of Size and Surface Area of Graphene Nanoplatelets on the Thermomechanical and Interfacial Properties of Shape Memory Multiscale Composites." *Polymer-Plastics Technology and Materials* 61 (12): 1–13. https://doi.org/10.1080/25740881.2022.2061864.

Tiwari, Nilesh, and AbdulHafiz A. Shaikh. 2022b. "Hybridization of Carbon Fiber Composites with Graphene Nanoplatelets to Enhance Interfacial Bonding and Thermomechanical Properties for Shape Memory Applications." *Polymer-Plastics Technology and Materials* 61 (2): 161–75. https://doi.org/10.1080/25740881.2021.1967390.

Trent, John S., Jeny I. Scheinbeim, and Peter R. Couchman. 1983. "Ruthenium Tetraoxide Staining of Polymers for Electron Microscopy." *Macromolecules* 16 (4): 589–98. https://doi.org/10.1021/ma00238a021.

Walrath, David E., and Donald F. Adams. 1983. "The Losipescu Shear Test as Applied to Composite Materials." *Experimental Mechanics* 23 (1): 105–10. https://doi.org/10.1007/BF02328688.

Wang, Mingtai, Lide Zhang, and Dezhu Ma. 1999. "Degree of Microphase Separation in Segmented Copolymers Based on Poly(Ethylene Oxide) and Poly(Ethylene Terephthalate)." *European Polymer Journal* 35 (7): 1335–43. https://doi.org/10.1016/S0014-3057(99)00030-0.

Webb, Graham A., and A. E. Aliew. 2006. *"Nuclear Magnetic Resonance. Chemical Society (UK)."* Royal Society of Chemistry: London.

Weller, Tanchum. 1977. "Experimental Studies of Graphite-Epoxy and Boron-Epoxy Angle Ply Laminates in Shear." NASA-CR-145231. https://ntrs.nasa.gov/citations/19780008156.

Whitney, James M., Isaac M. Daniel, and R. Byron Pipes. 1982. *Experimental Mechanics of Fiber Reinforced Composite Materials*. Society for Experimental Mechanics: Bethel, CT.

Yakacki, Christopher Michael, Robin Shandas, Craig Lanning, Bryan Rech, Alex Eckstein, and Ken Gall. 2007. "Unconstrained Recovery Characterization of Shape-Memory Polymer Networks for Cardiovascular Applications." *Biomaterials* 28 (14): 2255–63. https://doi.org/10.1016/j.biomaterials.2007.01.030.

Yeow, Y. T., and H. F. Brinson. 1978. "A Comparison of Simple Shear Characterization Methods for Composite Laminates." *Composites* 9 (1): 49–55. https://doi.org/10.1016/0010-4361(78)90519-0.

Zhang, Heng, Haitao Wang, Wei Zhong, and Qiangguo Du. 2009. "A Novel Type of Shape Memory Polymer Blend and the Shape Memory Mechanism." *Polymer* 50 (6): 1596–601. https://doi.org/10.1016/j.polymer.2009.01.011.

Zhang, Weijun, Meilin Yu, Yongqiang Cao, Zihan Zhuang, Kunxi Zhang, Dong Chen, Wenguang Liu, and Jingbo Yin. 2023. "An Anti-Bacterial Porous Shape Memory Self-Adaptive Stiffened Polymer for Alveolar Bone Regeneration after Tooth Extraction." *Bioactive Materials* 21 (March): 450–63. https://doi.org/10.1016/j.bioactmat.2022.08.030.

4 Modeling of Shape Memory Behavior

Shape memory polymers (SMPs) have limitations due to their lesser driving force as well as their thermoviscoelasticity, which has led to the development of shape memory polymer composites (SMPCs) (Leng et al. 2011). SMPCs, also known as shape memory nanocomposites, can significantly enhance mechanical properties by reinforcing SMPs with nanoparticles (zero-dimensional, such as Fe_3O_4), one-dimensional short fibers, continuous unidirectional fibers, carbon nanotubes (CNTs), or two-dimensional fiber fabrics and nano-paper as reinforcing phases. Increasing the volume fraction of the reinforcing phase enhances the mechanical and thermomechanical properties of SMPs and offers different driving techniques such as electricity (Lan et al. 2018), magnetism (Yu et al. 2013), light (Wang et al. 2022), and radio frequency (Li, Liu, and Leng 2015). These materials have several advantages, including high malleability, various driving techniques, adjustable transition temperature, low impact during deployment, and biocompatibility, making them suitable for diverse fields such as aerospace, flexible electronics, and bioengineering.

4.1 CONSTITUTIVE MODELS OF SMPs

To study various parameters like deformation behavior and cycle mechanism of SMPs, researchers have developed several constitutive models that can describe the material's complex thermoviscoelastic behavior. Currently, three different types of SMP constitutive models exist:

1. The rheological techniques that are based on the theory of viscoelasticity.
2. Theory of phase transformation that explains the thermomechanical nature of SMPs at a mesoscopic level.
3. The theory combination that integrates the two models mentioned above.

4.1.1 THEORY OF VISCOELASTICITY

The physical perspective, also known as thermoviscoelastic or rheological modeling, is a one class of SMP model that considers intrinsic properties of materials. These properties include cross-linking, molecular chain movements, contact surface movements, relaxation time, and intermolecular chain interactions. The kind of models that use this approach incorporate microstructural concepts and have a high level of accuracy in describing SMP behavior. This makes them suitable for modeling multi-SMPs, also called two-way SMPs. By considering the underlying polymer physics in amorphous SMPs, this approach allows for the inclusion of other phenomena such as aging or solution-driven shape memory effects in the creation of constitutive models (Zhao, Qi, and Xie 2015). Two important parameters considered in this approach are mobility and relaxation. To introduce modeling from this perspective, we will describe the standard linear solid (SLS) rheological model illustrated in Figure 4.1.

At temperatures above a certain point (T_{trans}), SMPs experience a significant reduction in viscosity, particularly in the damper. As a result, when a dashpot is combined with the spring in the Maxwell branch and loaded, the displacement of the dashpot is minimal, and most of the equilibrium branch's stored energy is entropic energy. If the temperature of SMP is lower under a constant strain, the viscosity of the damper increases significantly. Upon unloading, there is a slight deformation in the nonequilibrium branch, which has a higher spring constant than the equilibrium branch. This deformation becomes temporarily fixed in the nonequilibrium branch. Finally, by increasing the temperature of material, the viscosity of the dashpot decreases, and the spring stored force, which is in series with the dashpot, causes the dashpot to return to its original position, thereby restoring the SMP to its initial shape (Yu et al. 2012).

In 1997, Tobushi et al. introduced the first model for SMPs using a viscoelastic macroscopic approach, as shown in Figure 4.2. Their model utilized a four-element linear model for thermoset polyester polyol series PU and added a slip element to carry irreversible strain at a transition temperature $T\backslash T_{trans}$ (Tobushi et al. 1997). Other researchers have since developed SMP constitutive models based on this concept, incorporating factors such as internal friction, molecular chain direction, and cross-link decoupling. The slip element allows for some of the irreversible strain to remain if the internal friction is overcome by force. Tobushi et al. also proposed a relationship between irreversible strain (ε_s), creep strain (ε_c), and a reference strain (ε_l) defined at low temperature T_{low}, which can be expressed as $\varepsilon_s = C(\varepsilon_c - \varepsilon_l)$, where C is a temperature-dependent parameter. At high temperatures, ε_s is small and ε_l is high, whereas the opposite is true at low temperatures. One-dimensional versions of this constitutive model have also been developed, as stated in Eq. 4.1.

$$\dot{\varepsilon} = \frac{\dot{\sigma}}{E} + \frac{\sigma}{\mu} - \frac{\varepsilon - \varepsilon_s}{\lambda} + \alpha\dot{T} \tag{4.1}$$

The material model parameters like creep and irreversible strains described are denoted as E, μ, λ, and α in Figure 4.2. For model calibration, multiple dynamic mechanical thermal analysis tests were performed at varying temperatures to determine the temperature-dependent elastic modulus. It was noticed that all coefficients were dependent on temperature except the thermal expansion coefficient.

Viscoelastic Component

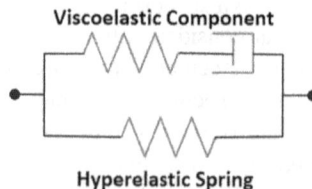

Hyperelastic Spring

FIGURE 4.1   SLS rheological model (Qi et al. 2008).

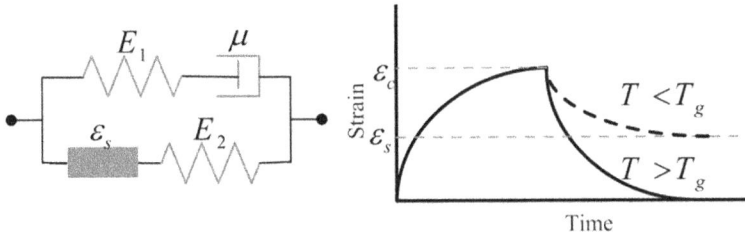

FIGURE 4.2 Tobushi et al.'s (1997) model with creep and creep recovery.

By conducting four dynamic mechanical thermal analysis tests, the creep and relaxation coefficients, as well as the thermal expansion coefficient, were identified. The model's predictions were verified using two thermomechanical stress and strain recovery tests. However, since rheologically linear viscoelastic elements were assumed in this model, its range of validity is limited to small strains. Additionally, the model lacked an explicit relationship between viscosity and relaxation time with respect to temperature, which was later modified by various researchers (Castro et al. 2010; Westbrook, Mather, et al. 2011; Nguyen et al. 2008). The Tobushi et al. (1997) model served as the basis for thermomechanical modeling of SMPs, and Lin and Chen (1999) derived a linear model and analyzed the behavior of an ester-type PU-based SMP based on the Tobushi et al. (1997) model, using two parallel branches of Maxwell to show a reversible and a fixed phase, as illustrated in Figure 4.3.

The model discussed in this study has a limitation in that it does not account for stress recovery, and it has been found to be inadequate in deriving the behavior of SMPs based on existing literature. However, Bhattacharyya and Tobushi (2000) created an analytical solution for the isothermal process using a four-element model previously presented by Tobushi et al. (1997). This model examined various parameters such as constant strain, constant stress, constant stress rate, constant strain, load periodic strain, and constant strain rate for the isothermal process on polyurethane (PU). It is worth noting that PU is not typically considered a constitutive model for SMP. Tobushi et al. (2001) later derived a nonlinear constitutive model for PU polyol series-based SMP using the same viscoelastic macroscopic approach but extended the model to large strains by incorporating a series of powers into their previous model. The governing equation for this new model is presented as Eq. 4.2.

$$\dot{\varepsilon} = \frac{\dot{\sigma}}{E} + m\left(\frac{\sigma - \sigma_y}{\mu}\right)^{m-1} + \frac{\sigma}{\mu} + \frac{1}{b}\left(\frac{\sigma}{\sigma_c}\right) - \frac{\varepsilon - \varepsilon_s}{\lambda} + \alpha\dot{T} \qquad (4.2)$$

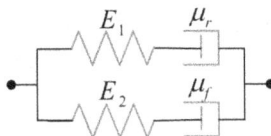

FIGURE 4.3 Lin and Chen's (1999) model for SMP.

The proportionality of stress in the viscous element, namely σ_c and σ_y, is independent of time in this model. To calibrate the linear model, six tests were performed in addition to a unidirectional tension test. Parameters such as m, n, a, and b are less influenced by temperature in comparison to other parameters. Therefore, these parameters were considered constant, and their evaluations were carried out as an average within the required temperature range.

Morshedian, Khonakdar, and Rasouli (2005) and Khonakdar et al. (2007) analyzed a three-element linear model (consisting of a spring and a couple of dashpots) and a four-element model (comprising a couple of springs, a sliding element, and one dashpot) for PE, respectively, based on the Tobushi et al. (1997) model. These models were developed for strains up to 200% under uniaxial loading. Bonner et al. (2010) derived a Kelvin–Voigt linear model for 35% weight of $CaCO_3$ with an amorphous lactide-based copolymer, which was calibrated using experiments but only examined the stress relaxation behavior of SMPs. Heuchel et al. (2010) utilized a rectified Maxwell model with two ways and the SLS model to examine the relaxation behavior of polyamide polyurethane in an isothermal process. Wong, Stachurski, and Venkatraman (2011) proposed a model for poly(d, l-lactide-co-glycolide) with a unidirectional concept, a biodegradable amorphous polymer similar to that of Morshedian, Khonakdar, and Rasouli (2005). Model studied for analysis of SMP behavior in the strain recovery step by considering a uni-dimensional flow rule and assuming the Arruda and Boyce (1993) eight-chain models for both the network and elastic elements in the equilibrium branch.

Li et al. (2015) developed a three-element rheological model for small 3D strains using the same technique as Tobushi et al. (1997). The model consisted of a Maxwell branch with a linear spring in series, and it was calibrated using the results of Tobushi et al. (1997). Shi et al. (2013), Balogun and Mo (2014), and Pan, Huang, and Liu (2019) also developed modified versions of the Tobushi et al. (1997) model in 3D, which were carried in ABAQUS using user material subroutines. Zhou, Liu, and Leng (2010) developed the 3D form of the constitutive equations by assuming the strain as an addition of three viscoelastic, thermal, and elastoplastic strains based on the theory of solid mechanics and viscoelasticity in the range of small strains. They performed tests on a thermoset styrene-based SMP and calibrated the model coefficients accordingly.

In a separate study, Ghosh and Srinivasa (2013) analyzed a one-dimensional thermoviscoelastic model for large strains based on the Tobushi et al. (1997) model using a Helmholtz potential and a four-element approach (considering a Kelvin element and a slip element parallel and a spring in series with this combination). The model was calibrated using Tobushi et al.'s (1997) conclusions on a PU-based SMP and Liu et al.'s (2006) conclusion on an epoxy resin to determine the model coefficients. It was not tested in stress recovery and is presented in state space form for two-network polymers, considering the yield stress of the SMP. The multi-network theory used in the model, which is proposed by Tobolsky and Andrews (1945) and Rajagopal and Wineman (1992), incorporates a friction dashpot element with yield stress 'k'.

Diani, Liu, and Gall (2006) proposed the pioneering nonlinear 3D model for SMP in the finite deformation range. Their model was designed for cross-linked SMP networks, specifically an epoxy thermoset resin. To calibrate the model coefficients, they relied on experimental results from Liu et al. (2006). As depicted in

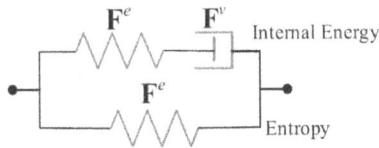

FIGURE 4.4   Rheological model of Diani, Liu, and Gall (2006) for SMP.

Figure 4.4, their model was based on the neo-Hookean model, which incorporated an entropic equilibrium branch as well as a nonequilibrium internal energy branch. The researchers reported the thermal expansion coefficient and elastic modulus for temperatures above and below T_g. However, they did not provide details on the theory behind the model or how it was calibrated.

Nguyen et al. (2008) presented a 3D model that is more comprehensive than previous models for the study of SMPs. The model was calibrated using experimental results from Qi et al. (2008) for SMP based on acrylate (tBA/PEGDMA). To capture the mechanical behavior of the SMP, they employed the theory of nonlinear viscoelasticity of Stefanie Reese and Govindjee (1998) and introduced the concepts of relaxation time as well as stress relaxation to model the movement of molecular chains because of the glassy phase transition. The model correlates the structural relaxation and viscoplastic flow below T_g, which is the glass transition temperature. Structural relaxation depends on the time at which molecular chains are reconfigured in the presence of a temperature change. The model did not consider the effects of heat conduction and pressure on structural relaxation or inelastic behavior. One of the important features of the model is the connection of the Adam–Gibbs model for an amorphous SMPs, which is a finite deformation thermo-elastic in nature. To analyze structural relaxation, they considered the Tool–Narayanaswamy–Moynihan model. The model also introduced fictive temperature through nonequilibrium behavior. Castro et al. (2010) applied the Kovacs–Aklonis–Hutchinson–Ramos (KAHR, 1979) model to account for the spectrum of relaxation time in amorphous polymers. In their model, they considered both structural relaxation and temperature-dependent viscoelasticity to investigate the stress–strain relationship. The model was only examined for 1D stress recovery in the small deformation range. Kohlrausch, Williams, and Watts method was used to calculate the stress relaxation. Theoretically and experimentally, the rate of temperature was investigated for the same SMP used by Qi et al. (2008) (tBA/PEGDMA) for the first time.

Yu et al. (2012) followed Castro et al.'s (2010) approach, which utilized the Kohlrausch, Williams, and Watts method along with a modified SLS model, to investigate the shape recovery behavior of an acrylate-based SMP (tBA/PEGDMA). In the meantime, Nguyen et al. (2010) improved their previous model by considering multiple Maxwell branches that become parallel to the equilibrium elastic branch for stress relaxation as well as a wider range of structural relaxation using KAHR theory. They analyzed the behavior of the SMP for various volume fractions of PEGDMA based on tBA during shape recovery. They utilized Haupt, Lion, and Backhaus (2000) method to generate relaxation spectra by using standard time–temperature superposition (TTSP) in their model. Choi et al. (2012) utilized Nguyen et al. (2010) model to study the effect of physical aging on an acrylate-based SMP (tBA/PEGDMA). Furthermore, Chen and Nguyen (2011) applied Nguyen et al.'s (2008, 2010) models to investigate the impact of various parameters, like cooling and heating rates, annealing time, strain rate, and temperature, on the thermomechanical nature of shape and stress recoveries for an acrylate-based SMP (tBA/PEGDMA).

Viscoelastic materials exhibit both strain-dependent as well as time-dependent responses. In this study, a hyperelastic model is utilized for the strain-dependent concept (i.e., the elastic part) due to the model's ability to accommodate large deformations. The strain energy function of neo-Hookean is adopted for the hyperplastic model. Typically, the viscoelastic theory(linear) can be described as Eq. 4.3 (Khajehsaeid et al. 2014):

$$\sigma(\varepsilon,t) \approx \int_{-\infty}^{t} \frac{d\sigma_0(\xi)}{d\xi} g(t-\xi).d\xi \tag{4.3}$$

The total stress, denoted by $\sigma(\varepsilon,t)$, can be expressed as the sum of the stress belonging to the response by the material, σ_0, and a dimensionless function represented by g. This function is often represented by the Prony series, which is given as Eq. 4.4:

$$g(\varepsilon,t) = g_\infty(\varepsilon) + \sum_{i=1}^{N} g_i(\varepsilon) \exp\left(\frac{-t}{\tau_i}\right) \tag{4.4}$$

The constants

g_∞ = belongs to equilibrium

g_i = instantaneous (viscous) parts

Both range between 0 and 1. It is also important to note that the sum of g_∞ and $\sum_{i=1}^{N} g_i$ must equal 1, where τ_i is the relaxation time in the ith branch, ranging from 1 to N.

The neo-Hookean model is utilized for the hyperelastic part of the model. For a homogeneous, isotropic, and incompressible elastomer, the Cauchy stress tensor is expressed as Eq. 4.5 (Holzapfel 2002):

$$\sigma^{hyper}(\varepsilon) = g_\infty(\varepsilon)\sigma_0(\varepsilon) = -pI + 2\left(\frac{\partial\psi}{\partial I_1} + I_1\frac{\partial\psi}{\partial I_2}\right)B - 2\frac{\partial\psi}{\partial I_2}B^2 \tag{4.5}$$

where $B = FF^T$ is the left Cauchy–Green deformation tensor whereas p = hydrostatic pressure. ψ is a strain energy function and it is a function of the invariants of B, represented by I_1 (i = 1, 2, 3) and defined as: $I_1 = tr(B)$, $I_2 = 1/2[(tr(B))^2 - tr(B^2)]$ and $I_3 = \det(B)$. ψ is calculated through $C_{10}(I_1 - 3)$ where C_{10} is a parameter regarding material. To account for temperature effects, the TTSP is utilized, which links temperature and timescale. In the above equation, t is replaced with t' to incorporate this phenomenon, stated as Eq. 4.6

$$t' = \int_0^t \frac{d\xi}{a(T)} \qquad (4.6)$$

The shift factor, denoted by $a(T)$, is a required parameter that needs to be determined in order to know the behavior of a material. It is particularly important to know the glass transition temperature (T_g) of the material in question. When the temperature is close to T_g, the shift factor can be calculated using the WLF equation, as shown in Eq. 4.7

$$\log(a(T)) = \frac{-C_1(T - T_r)}{C_2 + (T - T_r)} \qquad (4.7)$$

which takes into account material constants C_1 and C_2, and a referred temperature (T_r). In other words, if the temperature is above T_g, the shift factor can be analyzed using the Arrhenius-type equation with the consideration of a constant C, given by Eq. 4.8.

$$\ln(a(T)) = -\frac{1}{C}\left(\frac{1}{T} - \frac{1}{T_r}\right) \qquad (4.8)$$

It is worth noting that the Arrieta, Diani, and Gilormini (2014) model provided results that are shown in Figure 4.5.

In a similar approach to Arrieta, Diani, and Gilormini (2014) model, Fan et al. (2018) utilized the same model to conduct relaxation tests at varying temperatures in order to maintain the model coefficients through an optimization process. Their study focused on an epoxy-based SMP, and the results were presented in the form of recovery of strain and stress.

Xiao et al. (2016) also utilized a model with a few parallel nonequilibrium branches to determine the strain recovery nature of thermoplastic poly(para-phenylene) under large strains. They used a polynomial function that was dependent on temperature to evaluate the shift factor.

Meanwhile, Gu et al. (2016) aimed to improve the Arrieta, Diani, and Gilormini (2014) models by optimizing various terms like storage and loss modulus as a function of temperature based on the results of dynamic mechanical thermal analysis tests

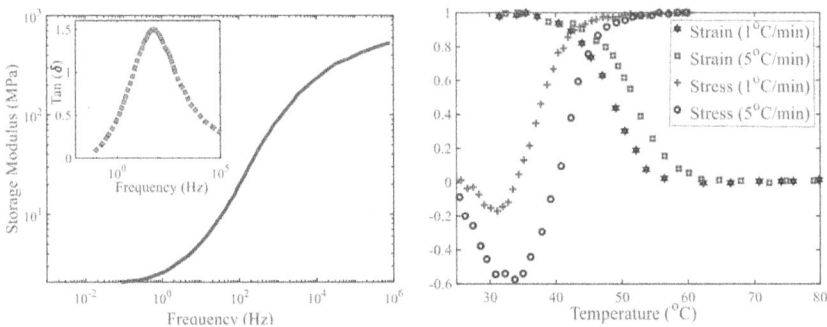

FIGURE 4.5   The experimental data from Arrieta, Diani, and Gilormini (2014) with storage modulus versus frequency relation.

conducted at a constant frequency. Through this approach, they can deduce viscoelastic coefficients and time–temperature–stress–pressure parameters simultaneously for a syntactic foam-based SMP and a styrene-based thermoset resin. Additionally, they obtained Arruda–Boyce coefficients by conducting a high-temperature tensile test. The model derived by Gu et al. (2016) has the ability to accurately replicate stress and strain recoveries.

Li and Liu (2017) developed a 3D finite strain rheological model for SMPs using a combination of springs and linear/nonlinear dashpots based on new rheological models for SMPs. The model was calibrated using experimental results from Tobushi et al. (1997, 2001) and McClung, Tandon, and Baur (2013) and verified in scenarios involving strain recovery for thermomechanical uniaxial and biaxial systems. Similarly, Saleeb, Natsheh, and Owusu-Danquah (2017) proposed a strain recovery of a thermoset PU in a 3D large strain model based on the viscoelastic approach, which shared similarities with their previous model for SMAs presented in Saleeb, Padula, and Kumar (2011). One advantage of this model for SMPs was the lack of material parameters required for calibration, and the researchers also found SMP behavior under repetitive loading. In contrast, Pulla et al. (2015) assumed SMPs to be hyperelastic materials and found model coefficients for various temperatures using a simple approach. Despite its simplicity, their model was able to identify the shape recovery response of an epoxy-based SMP.

Several proposed models for SMPs have utilized fractional viscoelastic constitutive models instead of classic viscoelastic models, which include a fractional dashpot. This approach is used to provide more accuracy in the results with a lower number of material parameters. To date, six models have been developed using this method: Sun and Gu (2016); Fang et al. (2018); Pan et al. (2018); Leng, et al. (2018); Xie, et al. (2018); and Fang, Sun, and Gu (2016). Sun and Gu (2016) used a three-element linear fractional-order Zener model and the viscoelastic fractional approach to predict the relaxation behavior of an acrylate-based SMP (tBA/PEGDMA), but did not account for the shape and force recovery responses. Similarly, Li et al. (2017) used a linear SLS model and the same fractional viscoelastic approach to study the shape recovery behavior of SMP.

Pan and Liu (2018) developed a 1D generalized Maxwell–Wiechert model along with two nonequilibrium subways to model the shape memory effect (SME) in a styrene-based thermoset polymer, but their model was only applicable for a small strain. Leng et al. (2018) applied a similar approach but used a generalized Maxwell–Wiechert model with four branches to describe the behavior of a Veriflex-epoxy as a thermoset SMP more accurately (based on experimental results reported by Yu et al. (2016) in shape regain). Xie et al. (2018) used a similar approach but applied the fractional Maxwell model in generalized form, which consists of a transient network composed of n parallel Maxwell branches in series with a backbone network, to account for multiple structural relaxations in 1D and successfully generate the shape and force regain in SMP. They utilized the KAHR model with 33 parameters for the thermal part and evaluated the model for results reported in an experimental way (Tobushi et al., 1997; Liu et al., 2006). Fang et al. (2018) also used this technique to examine multi-SMP. The rheological representations of the derived conventional SMP models from the viscoelastic ways are illustrated in Table 4.1, which represents the parts of the models from a mechanical point of view and is also presented simply due to its simplicity.

TABLE 4.1
Different Viscoelastic Models for SMP

Constitutive Models	Reference
	Li, He, and Liu 2017
	Lin and Chen 1999
Intermolecular Resistance / Molecular Network Resistance	Srivastava, Chester, and Anand 2010
	Wong, Stachurski, and Venkatraman 2011
	Ghosh and Srinivasa 2011
	Morshedian, Khonakdar, and Rasouli 2005
	Abrahamson et al. 2003

(Continued)

TABLE 4.1 (*Continued*)
Different Viscoelastic Models for SMP

Constitutive Models	Reference
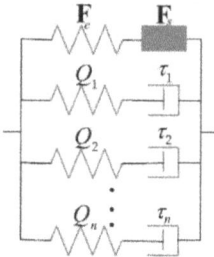	Bouaziz, Roger, and Prashantha 2017
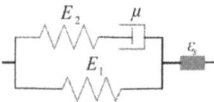	Balogun and Mo 2014
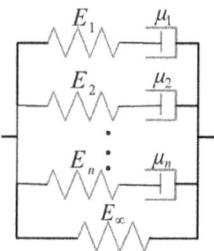	Pan and Liu 2018
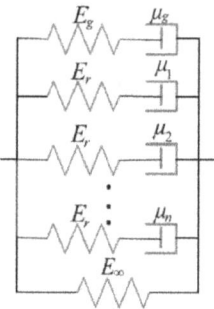	Westbrook, Kao, et al. 2011
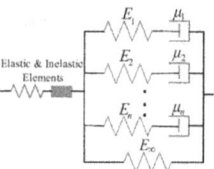	Gu, Sun, and Fang 2014
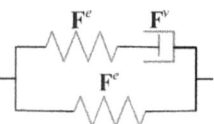	Nguyen et al. 2008

(Continued)

TABLE 4.1 (*Continued*)
Different Viscoelastic Models for SMP

Constitutive Models	Reference
	Tobushi et al. 1997
	Zeng, Xie, et al. 2018

SMP, shape memory polymer.

4.1.2 THE PHASE TRANSFORMATION THEORY

The SMP modeling approach based on the transition of phase theory assumes the material as a mixing of two distinct phases, namely active (rubbery) and frozen (glassy) phases, which describe how the material undergoes a phase transition with changes in temperature by considering a frozen phase volume fraction (Nguyen, 2013). This approach, in contrast to the viscoelastic approach, relies on observed phenomena and utilizes appropriate models and terms to simulate SMP behavior. Typically, this approach is more suitable for semicrystalline SMPs with distinct phases and is often based on physical concepts (Barot and Rao 2006; Barot, Rao, and Rajagopal 2008; Park et al. 2016; Reese and Christ 2010; Westbrook et al. 2010; Gu, Leng, and Sun 2017). While this approach can also be applied to amorphous SMPs, these materials do not undergo a true phase transition, and their models are generally considered from a phenomenological perspective. Few authors consider the concept of phase change a macro-perspective, but this study does not differentiate SMP models from a micro- or macro-modeling perspective and simply divides them into phase change, i.e., transition, and viscoelastic approaches. In this consideration, the SME is specified by parameters such as frozen phase volume fraction and stored strain, which provide a non-physical description of the phase change phenomenon (Abishera, Velmurugan, and Gopal 2017; Li and Xu 2011). Therefore, this approach does not consider the microscopic thermoviscoelastic phenomena that occur during the transition phase.

While viscoelastic models focus on the fundamental mechanisms responsible for shape memory effects (SMEs) by considering the dependence of molecular mobility on temperature and time, the phase transition approach models SMP behavior macroscopically, according to studies by Boatti, Scalet, and Auricchio (2016) and Luo et al. (2014). Boatti, Scalet, and Auricchio (2016) suggest that the phase transition perspective is more adaptable in modeling the behavior of SMPs, and the model coefficients can be obtained more easily and with lower computational costs, which may make it more appealing to engineers. In the following sections, we will explore the phase transition-based constitutive models found in the literature after providing this brief introduction to the phase transition approach.

A phenomenological constitutive model for amorphous SMPs was proposed by Liu et al. (2006) based on the phase transition approach. This model, which is considered one of the most fundamental models in the field, introduced the concept of stored strain and employed two internal variables—the stored strain and the frozen phase volume fraction—to explore the nature of transition in the microstructure. The model considers two active and frozen phases, where the transition between these phases contributes to the improvement of stored strain and the recovery mechanism (as depicted in Figure 4.6). It assumes that the stresses in both the frozen and glassy phases are the same. This model serves as the foundation for many subsequent models.

Liu et al.'s (2006) model has several limitations and assumptions, such as not considering time-dependent effects, using the Reuss phase composition by approximation for homogenization, assuming SMEs as a specific elastic problem, and proposing the model for 1D small strain deformations. Additionally, the evolution law is only shown for the cooling process, with no separate equation for the heating stage. Furthermore, the volume fraction of each phase is determined empirically using experiments. Specifically, in this model, the volume fraction determined for the frozen and active phases is defined as Eq. 4.9:

$$\varphi_a = \frac{V_a}{V}; \quad \varphi_f = \frac{V_f}{V}; \quad \varphi_a + \varphi_f = 1 \qquad (4.9)$$

This model defines the volume fraction of the active and frozen phases as φ_a and φ_f, respectively,

while V = total volume
V_a = active phase volume
V_f = frozen phase volume

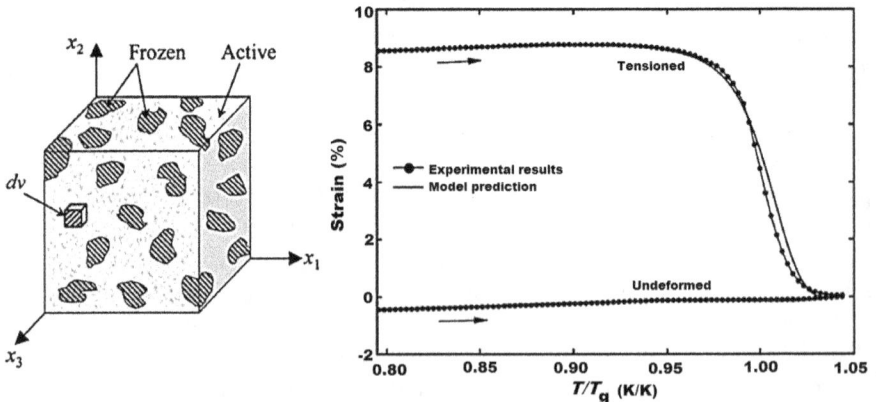

FIGURE 4.6 Scheme of the microstructure of SMP and recovery strains with their experimental validation (Liu et al. 2006).

The model assumes that the strain (ε) consists of three components, namely the stored strain (ε_s), elastic mechanical strain (ε_m), and thermal strain (ε_T), represented by Eq. 4.10:

$$\varepsilon = \varepsilon_s + \varepsilon_m + \varepsilon_T \tag{4.10}$$

For the cooling process the stored strain is calculated using Eq. 4.11

$$\varepsilon_s = \int_0^{\varphi_f} \varepsilon_f^e d\varphi \tag{4.11}$$

where ε_f^e = entropic frozen strain.

The mechanical elastic strain is calculated using Eq. 4.12:

$$\varepsilon_m = \left[\varphi_f S_i + \left(1 - \varphi_f\right) S_e \right] : \sigma \tag{4.12}$$

where σ is the stress tensor and the double contraction of tensors denoted by ":". It is assumed that the stress is the same in both the frozen and active phases. The fourth-order compliance tensor S_i is linked to the internal deformation energy, while the compliance tensor S_e shows the entropic deformation. The thermal strain is determined by Eq. 4.13:

$$\varepsilon_T = \left[\int_{T_0}^T \left\{ \varphi_f \alpha_f (\theta) + \left(1 - \varphi_f\right) \alpha_a (\theta) d\theta \right\} \right] I \tag{4.13}$$

where α_a and α_f are the thermal expansion coefficients for the active as well as frozen phases, respectively, and I is the identity tensor. Based on this model, the strain-strain relation is given by Eq. 4.14:

$$\sigma = \left[\varphi_f S_i + \left(1 - \varphi_f\right) S_e \right]^{-1} : \left(\varepsilon - \varepsilon_s - \varepsilon_T\right) \tag{4.14}$$

Figure 4.6 depicts the strain recovery prediction of the Liu et al. (2006) model for the SMP during the heating process, compared with experimental data. However, this model was presented in 1D and did not provide any constitutive relations for the heating stage. To overcome this limitation, Chen and Lagoudas (2008a) extended Liu et al.'s (2006) model by introducing a 3D finite strain model based on nonlinear thermo-elastic theory that neglects hysteresis. They employed the neo-Hookean model to account for maximum deformations and used the Reuss approximation for homogenization of different phases. Finally, they calibrated the model coefficients using the experimental data from the investigation of an epoxy resin noted by Liu et al. (2006) and correlated them in two stress and strain recovery paths, which yielded more accurate results.

Chen and Lagoudas (2008b) derived a linearized version of the SMP model based on the assumption of small deformations. Meanwhile, Volk et al. (2010a) conducted experimental tests on a Veriflex-based SMP and used the model proposed by Chen and Lagoudas (2008a, 2008b) after calibrating it. Qi et al. (2008) introduced a new function for the frozen phase volume fraction and employed a Voigt-type assumption for homogenization to develop a finite strain constitutive model. They applied this model to an acrylate-based SMP to regenerate strain and stress recovery paths. In their experiments, they divided the SMP into three phases and developed constitutive equations for each of them. Wang et al. (2009) argued that Liu et al.'s (2006) model has some limitations and proposed a new 1D model that accounts for rate-dependent effects. They derived a new rate-dependent function for the frozen phase volume fraction.

A new constitutive model was developed by Pan et al. (2018) based on Liu et al.'s (2006) model, taking into account a rate-dependent T_{trans}. The model was calibrated for polystyrene and an acrylate-based SMP (tBA/PEGDMA) using test results from Arrieta, Diani, and Gilormini (2014). Similarly, Guo et al. (2017) proposed a different function for the frozen phase volume fraction and used a three-element model for the frozen phase and a four-element model for the active phase. They successfully modeled and calibrated experimental results from Liu et al. (2006) and Tobushi et al. (1997) to detect the strain recovery stage of SMPs.

Gilormini and Diani (2012) developed a different constitutive model of SMPs by modifying existing models proposed by Liu et al. (2006), Chen and Lagoudas (2008a), and Qi et al. (2008) and introducing new functions for the frozen phase volume fraction. They examined the accuracy of various similar approximations, including Reuss, Voigt type, Hashin–Shtrikman upper and lower bounds, and self-consistent models, to predict the shape memory effect of SMPs using the experimental results of Liu et al. (2006). They found that the Voigt model had the highest accuracy.

Kim, Kang, and Yu (2010) proposed a 3D finite strain model for SMPs, assuming that the material comprises three phases: a frozen phase, an active phase, and a hard phase. They used two hyperelastic soft-phase models (Mooney–Rivlin) and a viscoelastic hard-phase model (SLS) to describe the mechanical behavior of each phase. Gu, Sun, and Fang (2015) also developed a 3D model for SMPs, but they employed the affine network model instead of the Mooney–Rivlin model for the two soft phases. They considered the microstructure changes during the shape recovery process and used a three-element rheological component in their model.

Xu and Li (2010) made improvements to Liu et al.'s (2006) model to analyze effects that are time-dependent and introduced a 3D model for thermodynamic processes of stress and strain recovery for syntactic foam. They considered the frozen phase volume fraction function proposed by Qi et al. (2008) and ensured the coefficients had physical meaning. Baghani et al. (2012) investigated a 3D model for a thermoset SMP under multi-axial thermomechanical loading, time-dependent, where strain was decomposed into six components (Figure 4.7). They satisfied the second law of thermodynamics in the form of the Clausius–Duhem inequality and accounted for time-dependent effects. They tested the model for experiments by Volk et al. (2010b); Volk, Lagoudas, and Chen (2010); and Li and Nettles (2010). The model, stated as Eq. 4.15, was implemented in FEM to calculate strain and analyze a 3D beam for the application as an SMP-based medical stent.

$$\varepsilon = \emptyset_p \varepsilon_p + \emptyset_h \varepsilon_h + \varepsilon_i + \varepsilon_T \qquad (4.15)$$

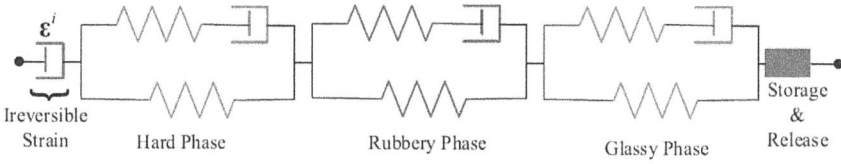

FIGURE 4.7   Phase transition model by Baghani et al. (2012).

The volume fractions of the hard segment and the SMP segment are denoted by $\varnothing_p$ and $\varnothing_h$, respectively, while the irreversible strain is represented by ε_i. The hard segment refers to the microballoons mixed in the material, which increase its mechanical strength. Figure 4.13 illustrates a detailed representation of this constitutive model. The strain of the SMP segment, ε_p, is decomposed as shown in Eq. 4.16.

$$\varepsilon_p = \varphi_r \varepsilon^r + \varphi_g \varepsilon^G \tag{4.16}$$

The equation in the glassy phase, total strain (ε^G) involves subscripts r and g which show the rubbery and glassy phases, respectively. The model assumes that $\varnothing_p$ and $\varnothing_h$ are constant parameters, while φ_r and φ_g are temperature-dependent. The expression for $\varphi_g \varepsilon^G$ can be derived using Eq. 4.17.

$$\varphi_g \varepsilon^G = \varphi_g \varepsilon^G + \varepsilon^{is} \tag{4.17}$$

$\varepsilon^{is} = \int \varepsilon^r d\varphi_g$ during the cooling and $\varepsilon^{is} = \int \varepsilon^r / d\varphi_g$ during the heating.

$$\dot{\varepsilon}^{is} = \varphi_g' \dot{T} \left(k_{s2} \frac{\varepsilon^{is}}{\varphi_g} \right) \tag{4.18}$$

Additionally, evolution laws are proposed for other internal variables,

$$\dot{\varepsilon}^{ir} = \frac{1}{\eta_r} \frac{\partial \psi_r^{neq}}{\partial \varepsilon^{er}}; \quad \dot{\varepsilon}^{ig} = \frac{1}{\eta_g} \frac{\partial \psi_g^{neq}}{\partial \varepsilon^{eg}}; \quad \dot{\varepsilon}^{ih} = \frac{1}{\eta_h} \frac{\partial \psi_h^{neq}}{\partial \varepsilon^{eh}}; \quad \dot{\varepsilon}^i = \frac{1}{\eta_i} \sigma \tag{4.19}$$

which involve the free energy density function (ψ) and the viscosity coefficient (η).

Baghani, Arghavani, and Naghdabadi (2014) developed a 3D mechanical model that relied on classical thermodynamic structures and utilized a logarithmic strain, which is a more accurate measure of strain. The deformation gradient tensor was split into two parts: elastic and stored, with the stored part consisting of rubbery and glassy components. Evolution equations for the internal variables were presented for both cooling and heating processes. Taherzadeh et al. (2016) employed the Baghani et al. (2012) model to examine the homogenization of SMP nanocomposites and numerically investigated large deformations of graphene-reinforced SMPs.

Guo et al. (2014) introduced a novel relation for the frozen phase volume fraction and developed a 3D small strain model based on phase transition techniques. This model successfully predicted the process for force and shape recovery of an amorphous styrene-based epoxy-type SMP. The model accounted for both temperature and stress as important factors in the phase transition and included elastic linear, three-element active, and strain storage elements in series, following the approach of Liu et al. (2006). The model also deployed a 3D version of the rheological stress–strain relationship proposed by Tobushi et al. (1997) to describe viscoelastic strain. In another work by Guo et al. (2015), the authors used a strain rate parameter to examine the epoxy-type SMP and its strain hardening behavior, and proposed a new model for the rubbery phase (RP) and generalized Kelvin elements for the frozen phase. Guo et al. (2016) used a new function for the volume fraction of the frozen phase to investigate the strain recovery behavior of SMP and proposed a new model for the RP and Kelvin elements for the frozen phase. From a thermoviscoelastic point of view, Guo et al. (2016) proposed equations for an SMP which is large deformation constitutive. Additionally, Zeng et al. (2019) used the concept of frozen phase volume fraction in combination with viscoplastic and hyperelastic stresses to investigate the thermomechanical behavior of a thermoset epoxy, following a phase transition approach.

Pieczyska et al. (2015) proposed a shape recovery process model of PU SMPs based on the phase transition approach and the volume fraction of the frozen phase given by Qi et al. (2008). Their model consisted of two parallel rubbery and glassy phases, with the RP modeled as a hyperelastic combination of Arruda–Boyce material and the glassy phase considered as a three-element hyperelastic–viscoplastic (Zener) model. The model was evaluated for the hysteresis behavior of SMPs and the impact of loading cycles.

Yang and Li (2016) derived a temperature- and time-dependent constitutive model for amorphous SMPs based on composite material theory. They considered the SMP as a composite material, with the matrix being the inactive phase and the dispersed particles doing the work of the active phase. They applied the Mori–Tanaka theory to obtain the effective mechanical properties of the SMP and proposed a new function for the volume fraction of the frozen phase. The coefficients were computed from experimental results, and the model was evaluated in both the strain and stress recovery processes, showing consistency with experimental results and physically meaningful terminology.

In 2016, Park and colleagues proposed a constitutive model for two-phase materials that can undergo large deformations under multi-axial loading up to about 200% strain (Park et al. 2016). Their model included a new function for the volume fraction of the frozen phase and utilized a three-element model for the rubbery part and a three-element Kelvin model for the glassy part. The deformation gradient was degraded into hyperelastic, viscoelastic, viscoplastic, and shape recovery components. The authors employed the Poynting-Thomson three-element model for the rubbery part and a specific rheological model with six elements for the frozen part. They developed the model by defining the Helmholtz free energy function and using the Clausius–Duhem inequality to consider the second law of thermodynamics. The model was tested for experiments on a SMP, but its high number of material coefficients made calibration challenging. This model only evaluates the strain recovery process. The same model incorporated residual stresses to simulate woven fabric-mixed SMP composites. They employed classical anisotropic hyperelastic theorems to evaluate the orthotropic characteristics of the woven fabric-reinforced SMP.

To calculate the total stress in SMPs, researchers have employed Eshelby's inclusion theorem to account for the stress generated by the constitutive equation, anisotropic hyperelastic model, and thermal residual stress. Li, Hu, and Liu (2017) took a different approach, splitting the rubbery, frozen phases into numerous sub-phases connected in series and assuming a linear elastic nature for each phase to investigate the shape and force recovery nature of SMPs. They based their model on experimental results from Liu et al. (2006) and Arrieta, Diani, and Gilormini (2014). In a subsequent work, Li and Liu (2018) investigated a strain model for SMP recovery processes by considering a three-element viscoelastic SLS model for the RP, an elastic spring for the glassy phase, and using the frozen phase volume fraction of Qi et al. (2008).

Dong et al. (2018) adopted a similar approach to Qi et al. (2008) and developed a nonlinear three-element model for the frozen and active phases of a PU-based SMP. Their equations were formulated in one dimension, and they tested their model by predicting the nature of the SMP during a strain recovery step. Meanwhile, Gu, Leng, and Sun (2017) took a slightly different approach and reviewed dependent internal state variables for an acrylate-based SMP (tBA/PEGDMA) from a phase change standpoint, using a three-element model. They incorporated the Arruda–Boyce model for the equilibrium part and a flow rule theory for the nonequilibrium part, taking into consideration structural and stress relaxation. To determine the SMP Poisson's ratio, they utilized the frozen phase volume fraction function developed by Qi et al. (2008) and validated the model with previous experimental data (Westbrook, Kao, et al. 2011).

Su and Peng (2018) adopted the approach of Reese, Böl, and Christ (2010) to decompose the free energy and proposed a 3D constitutive model for shape recovery. Their model included a three-element viscoelastic model and a flow rule theory similar to Nguyen et al.'s (2008) model. They also assumed a fictive temperature as an internal variable and defined material parameters using the phase transition and frozen volume fraction concepts. Table 4.2 summarizes the different forms of the frozen phase volume fraction available in the literature.

Some researchers have shown interest in semicrystalline SMPs due to their real phase transition (Barot and Rao 2006; Barot, Rao, and Rajagopal 2008; Ge, Yu, et al. 2012; Kolesov, Dolynchuk, and Radusch 2013). Barot and Rao (2006) and Barot, Rao, and Rajagopal (2008) developed a thermomechanical model for semicrystalline SMPs. Their model divided the SMP into two solid crystalline phases and a RP, treated in parallel, and combined the amorphous and crystalline energies to form the energy density function. The resulting equations of stress–strain–temperature were used in a 3D shape-retrieval process. Ge, Luo et al. (2012) investigated a finite strain 3D model for an SMP composite, prepared by mixing an elastomeric matrix and a semicrystalline fiber PCL, using the rule of mixture and accounting for stress concentration. They employed Avrami's theory to incorporate the crystal-melt phase transition and derived results for two thermomechanical stress and recovery processes for strain. Ge et al. (2016) later analyzed a production technique and constitutive model for anisotropic SMPs using a finite strain 3D model that connected the phase evolution theory for active soft materials to the composite theory with anisotropic thermal strains, enabling the regeneration of experiments for shape recovery.

TABLE 4.2
Some Prominent Correlations for Phase Transition Approach

Governing Equation	References
$\varphi(T) = 1 - \dfrac{1}{1 + c_f (T_h - T)^n}$	Liu et al. 2006
$\varphi(T) = \alpha \exp\left(-\left(\dfrac{T_t}{T}\right)^m \beta^n \right)$	Wang et al. 2009
$\varphi(T) = \dfrac{1}{1 + \exp\left(\dfrac{2w}{T - T_t}\right)}$	Reese, Böl, and Christ 2010
$\varphi(T) = \dfrac{\tanh\left(\dfrac{T_{\max} - A}{B}\right) - \tanh\left(\dfrac{T - A}{B}\right)}{\tanh\left(\dfrac{T_{\max} - A}{B}\right) - \tanh\left(\dfrac{T_{\min} - A}{B}\right)}$	Volk, Lagoudas, and Maitland 2011; 2011
$\varphi(T) = \left[1 - \left(\dfrac{T - T_{\min}}{T_{\max} - T_{\min}}\right)^m\right]^n$	Gilormini and Diani 2012
$\varphi\left(T, \dot{T}\right) = 1 - \displaystyle\int_{r_c(T)}^{\infty} p(r)\,dr \times \left[1 - \left\{1 - \exp\left(-\dfrac{nH_a(T)}{k_B T}\right)\right\}^{\frac{nt}{r_0}}\right]$	Yang and Li 2016
$\varphi(T) = \dfrac{1}{1 + \exp\left(-\dfrac{(T - T_t)}{A}\right)}$	Qi et al. 2008
$\varphi_{sfo} = \varphi_{sf} - \displaystyle\int_t \dfrac{d\varphi_{sa}}{dt}\,dt$	Kim, Kang, and Yu 2010
$\varphi(T) = \displaystyle\int_{T_s}^{T} \dfrac{1}{S\sqrt{2\pi}} \exp\left(-\dfrac{T - T_g}{2S^2}\right) dT$	Guo et al. 2016
$\varphi_g = \begin{cases} 1 & 1 < \varphi_g^0 \\ \varphi_g^0 & 0 \le \varphi_g^0 \le 1 \\ 0 & \varphi_g^0 \le 1 \end{cases} \qquad \varphi_g^0 = \dfrac{b}{1 + \exp\left(c(T - T_{tr})\right)} - d$	Park et al. 2016
$1 - \varphi_f = \varphi \exp\left(-\left(\dfrac{kT_{tran}}{T - \tau\beta}\right)^m \middle/ \beta^n \right)$	Guo et al. 2017

(Continued)

TABLE 4.2 (Continued)

Some Prominent Correlations for Phase Transition Approach

Governing Equation	References

$$\varphi\left(T,\dot{T}\right)=1-\frac{1}{1+\exp\left[-\left\{T-T_{tr}\left(\dot{T}\right)/b\right\}\right]}$$

Pan et al. 2018

$$\varphi_f=1-\gamma=1-AT\,\exp\left[-\frac{\Delta G\left(T_h\right)10^{\frac{C_1(T-T_h)}{C_1+T-T_h}}}{RT}+\frac{T_h-T}{bT_h-T}\right]$$

Lu et al. 2018

$$\varphi_f=1-\frac{1-b}{1+\exp\left(-a_1\left(T-T_r\right)\right)}-\frac{b}{1+\exp\left(-a_2\left(T-T_c\right)\right)}$$

Bouaziz, Roger, and Prashantha 2017

$$\varphi=1-\frac{1}{1+\exp\left(-\frac{T-T_{tr}\left(\dot{T}\right)}{b}\right)}$$

Pan et al. 2018

$$\varphi=1+\frac{\tanh\left(\gamma_1T_g-\gamma_2T\right)-\tanh\left(\gamma_1T_g-\gamma_2T_h\right)}{\tanh\left(\gamma_1T_g-\gamma_2T_h\right)-\tanh\left(\gamma_1T_g-\gamma_2T_1\right)}$$

Mostafa Baghani et al. 2012

$$\varphi=\frac{1}{V}\int_{\Omega}\tfrac{1}{2}\left(1+\tanh\left(\psi/s\right)dV\right)$$

Arvanitakis 2019

In a different study, Bouaziz, Roger, and Prashantha (2017) proposed a 3D model for semicrystalline thermoplastic PU in various recovery scenarios at large strains. They incorporated a rheological model, similar to the generalized Maxwell model with n nonequilibrium branches, to account for the viscoplastic behavior of SMP. Their model also included an equilibrium branch with a spring and slipping element to represent hyperelastic and inelastic behavior, respectively. They introduced a new function for the frozen phase volume fraction and utilized the finite strain viscoelastic relations in their constitutive relations.

4.1.3 THE COMBINATION THEORY

The rheological theory is effective in explaining stress relaxation, but it doesn't provide a complete understanding of how materials store and release strain, which the phase transition theory can explain. However, the phase transition theory doesn't fully account for the time–temperature equivalence of polymers. To address this, Qi et al. (2008) proposed a three-phase-transition constitutive model that combines phase transformation theory and viscoelasticity theory. This model explains the viscoelasticity and rate dependence of polymers and connects the shape memory effect (SME) to the glass transition. The three phases in this model are the RP above T_g, the frozen glassy phase resulting from cooling-induced RP conversion, and the initial glassy phase.

Baghani et al. (2012) created a 3D model that considered the material's viscous effects and confirmed its precision when subjected to time-dependent thermomechanical loads. In a similar manner, Kim et al. devised a three-phase phenomenological model for shape memory polyurethanes that employed a viscoelastic model to depict the hard-segment phase and a Mooney–Rivlin model for the active and frozen soft segments. Guo et al. (2016) introduced a new constitutive model based on viscoelasticity and phase transition theories, which included a normal distribution model that had fewer parameters and a more reasonable physical meaning. Park et al. (2016) proposed a constitutive model that broke down the overall deformation gradient multiplication into hyperelasticity, visco-elasticity, viscoplasticity, and shape memory strain, allowing it to describe multi-axial loading and significant deformations of up to 200% strain under the assumption that the shape memory strain was proportionate to the total deformation.

During the heating or cooling of a polymer, the transfer of temperature happens gradually from the surface to the interior. Similarly, the transformation of the material's phase should also take place gradually, from the surface toward the inside. However, it is not reasonable to assume that the frozen and active phases are uniformly distributed within the polymer and that the transition between the two phases occurs simultaneously, based on the microstructures of particle-reinforced composites and unidirectional fiber-reinforced polymer composites. To address this issue, Guo et al. (2017) proposed a novel microstructure for a semicrystalline SMP, which includes an active phase with a constant network and a frozen phase that can switch between a free and frozen state.

A constitutive model was developed by Liu et al. to overcome the limitations of previous models, which were developed only for specific materials. The new model uses the multiplicative decomposition of deformation gradients and is capable of accurately describing large deformations of various types of SMPs. One of the unique features of this model is its ability to consider phase transitions that occur due to the sudden formation and disappearance of reversible phases, as well as transitions from one phase to another (Li, He, and Liu 2017). Furthermore, Li and Liu (2018) established a new constitutive model that can transform into pure elasticity when the material's viscosity is not taken into account.

4.2   MICROMECHANICS OF THREE-PHASE COMPOSITES

This section focuses on and delves into the micromechanics of three-phase multiscale nanocomposites, which are composite materials consisting of fiber-reinforced layers with enhanced mechanical properties due to the addition of CNTs to the polymer matrix (epoxy). The composite is composed of three phases: the polymer matrix, the E-glass reinforcing fibers, and the arbitrarily oriented CNTs, collectively requiring a thorough mechanical description to evaluate the overall mechanical properties of the composite. Initially, the Eshelby–Mori–Tanaka scheme is utilized to estimate the mechanical properties of the CNT-enriched matrix. Subsequently, the oriented reinforcing fibers are integrated into the hybrid matrix, necessitating a suitable homogenization technique (HT) to obtain a comprehensive mechanical representation of the composite, which is transversely isotropic. Various techniques proposed by Hashin and Rosen, Halpin and Tsai, Chamis, and Hahn are discussed in this section.

4.2.1 Equivalent Continuum Model for CNTs

Assuming that the solitary nanofiber is a cylindrical solid that is linear elastic and homogeneous, we can consider it to be transversely isotropic, with its plane of isotropy transverse to its longitudinal axis, as stated in the work by Odegard et al. (2003) and supported by various molecular models. To fully define the constitutive equations of this equivalent continuum model, five independent material properties, represented by the elastic moduli C_{ij}^r, are needed. Hashin and Rosen (1964) proposed a way to express the relationships between stresses and strains for a transversely isotropic medium in terms of these five elastic moduli, as indicated by Eq. 4.20.

$$
\begin{bmatrix} \sigma_{11}^r \\ \sigma_{22}^r \\ \tau_{12}^r \\ \tau_{13}^r \\ \tau_{23}^r \\ \sigma_{33}^r \end{bmatrix} = \begin{bmatrix} C_{11}^r & C_{12}^r & 0 & 0 & 0 & C_{12}^r \\ C_{12}^r & C_{22}^r & 0 & 0 & 0 & C_{23}^r \\ 0 & 0 & C_{44}^r & 0 & 0 & 0 \\ 0 & 0 & 0 & C_{44}^r & 0 & 0 \\ 0 & 0 & 0 & 0 & \frac{C_{22}^r - C_{23}^r}{2} & 0 \\ C_{12}^r & C_{23}^r & 0 & 0 & 0 & C_{22}^r \end{bmatrix} \begin{bmatrix} \varepsilon_{11}^r \\ \varepsilon_{22}^r \\ \gamma_{12}^r \\ \gamma_{13}^r \\ \gamma_{23}^r \\ \varepsilon_{33}^r \end{bmatrix} \tag{4.20}
$$

The stress and strain components of the fiber in the local reference system $\hat{x}_1 \hat{x}_2 \hat{x}_3$ (as shown in Figure 4.8) are represented by σ_{11}^r, σ_{22}^r, τ_{12}^r, τ_{13}^r, τ_{23}^r, σ_{22}^r and ε_{11}^r, ε_{22}^r, γ_{12}^r, γ_{13}^r, γ_{23}^r, ε_{22}^r. A more convenient way to express the stiffness matrix can be achieved by adopting the notation suggested by Hill (1964a, 1964b) in Eq. 4.21.

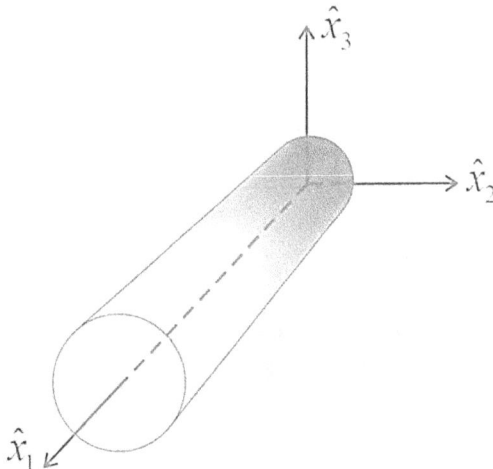

FIGURE 4.8 Stress and strain components of the fiber in the local reference system.

$$
\begin{bmatrix} \sigma_{11}^r \\ \sigma_{22}^r \\ \tau_{12}^r \\ \tau_{13}^r \\ \tau_{23}^r \\ \sigma_{33}^r \end{bmatrix} = \begin{bmatrix} n_r & l_r & 0 & 0 & 0 & l_r \\ l_r & k_r+m_r & 0 & 0 & 0 & k_r-m_r \\ 0 & 0 & p_r & 0 & 0 & 0 \\ 0 & 0 & 0 & p_r & 0 & 0 \\ 0 & 0 & 0 & 0 & m_r & 0 \\ l_r & k_r-m_r & 0 & 0 & 0 & k_r+m_r \end{bmatrix} \begin{bmatrix} \varepsilon_{11}^r \\ \varepsilon_{22}^r \\ \gamma_{12}^r \\ \gamma_{13}^r \\ \gamma_{23}^r \\ \varepsilon_{33}^r \end{bmatrix}
\tag{4.21}
$$

The five autonomous constants recognized as Hill's elastic moduli, denoted by k_r, l_r, m_r, n_r and p_r, can be used to define the stiffness matrix. Table 4.3 provides the Hill's elastic moduli for the most common types of single-walled carbon nanotubes, which can be categorized based on their chiral indices, represented by the symbols n and m that describe the arrangement of the carbon atoms. For the purposes of this paper, only armchair nanotubes are being considered, so we can assume that n is equal to m. The relationship among the elastic moduli C_{ij}^r and Hill's parameters is evidently defined in the technical manuscript by Halpin (1969) as Eqs. 4.22–4.26.

$$
k_r = \frac{C_{22}^r + C_{23}^r}{2}
\tag{4.22}
$$

$$
l_r = C_{12}^r
\tag{4.23}
$$

$$
m_r = \frac{C_{22}^r - C_{23}^r}{2}
\tag{4.24}
$$

$$
n_r = C_{11}^r
\tag{4.25}
$$

$$
p_r = C_{44}^r
\tag{4.26}
$$

TABLE 4.3

Mechanical Properties of SWCNT (Tornabene et al. 2016)

CNTs	k_r (GPa)	l_r (GPa)	m_r (GPa)	n_r (GPa)	p_r (GPa)
SWCNT(5, 5)	536	184	132	2143	791
SWCNT(10, 10)	271	88	17	1089	442
SWCNT(15, 15)	181	58	5	726	301
SWCNT(20, 20)	136	43	2	545	227
SWCNT(50, 50)	55	17	0.1	218	92

Similarly, the mechanical performance of the equivalent continuum model can be defined using the engineering constants E_1^r, E_2^r, E_3^r, v_{12}^r, v_{13}^r, G_{12}^r, G_{13}^r and G_{23}^r, as defined by Eqs. 4.27–4.31.

$$E_1^r = n_r - \frac{l_r^2}{k_r} = C_{11}^r - \frac{2\left(C_{12}^r\right)^2}{C_{22}^r + C_{23}^r} \tag{4.27}$$

$$E_2^r = E_3^r = \frac{4m_r\left(k_r n_r - l_r^2\right)}{k_r n_r - l_r^2 + m_r n_r} = \frac{\left(C_{22}^r - C_{23}^r\right)\left(C_{11}^r C_{22}^r + C_{11}^r C_{23}^r - 2\left(C_{12}^r\right)^2\right)}{C_{11}^r C_{22}^r - \left(C_{12}^r\right)^2} \tag{4.28}$$

$$v_{12}^r = v_{13}^r = \frac{l_r}{2k_r} = \frac{C_{12}^r}{C_{22}^r + C_{23}^r} \tag{4.29}$$

$$v_{23}^r = \frac{n_r\left(k_r - m_r\right) - l_r^2}{n_r\left(k_r + m_r\right) - l_r^2} = \frac{C_{11}^r C_{23}^r - \left(C_{12}^r\right)^2}{C_{11}^r C_{22}^r - \left(C_{12}^r\right)^2} \tag{4.30}$$

$$G_{12}^r = G_{13}^r = p_r = C_{44}^r \tag{4.31}$$

$$G_{23}^r = m_r = \frac{C_2^r}{2\left(1 + v_{23}^r\right)} \tag{4.32}$$

Note that in Eqs. 4.27–4.32, five independent measurements arise from the fiber's transverse isotropy, and that the other engineering variables need solving for $i, j = 1$, 2, and 3, as shown in Eq. 4.33.

$$\frac{v_{ij}^r}{E_i^r} = \frac{v_{ji}^r}{E_j^r} \quad G_{ij}^r = G_{ji}^r \tag{4.33}$$

For transversely isotropic media, Hill's modulus can be expressed as a function of the fiber engineering coefficients, given by Eqs. 4.34–4.36.

$$k_r = \frac{E_2^r}{2\left(1 - v_{23}^r - 2v_{21}^r v_{12}^r\right)} \tag{4.34}$$

$$l_r = \frac{v_{12}^r E_2^r}{\left(1 - v_{23}^r - 2v_{21}^r v_{12}^r\right)} = 2v_{12}^r k_r \tag{4.35}$$

$$m_r = \frac{E_2^r}{2\left(1 + v_{23}^r\right)} = \frac{1 - v_{23}^r - 2v_{21}^r v_{12}^r}{1 + v_{23}^r} k_r \tag{4.36}$$

The defined relationship over here can also be useful in evaluating the effective properties of the fiber-reinforced composites.

$$n_r = \frac{E_1^r \left(1 - v_{23}^r\right)}{\left(1 - v_{23}^r - 2v_{12}^r v_{21}^r\right)} = 2\left(1 - v_{23}^r\right)\frac{E_1^r}{E_2^r}k_r \tag{4.37}$$

$$p_r = G_{12}^r \tag{4.38}$$

4.2.2 MECHANICAL CHARACTERIZATION OF A MATRIX EMBEDDED WITH CNTs

CNTs can be utilized as nanofillers to improve the overall properties of a matrix, such as epoxy resin. In this scenario, the matrix is assumed to be isotropic and categorized by Young's modulus (E_m) and Poisson's ratio (μ_m). Eshelby (1957) and Mori and Tanaka (1973) schemes are utilized to estimate the mechanical properties of the resulting hybrid matrix enhanced by CNTs. While simpler arrangements could be considered, the Mori–Tanaka approach enables modeling of additional advantages, such as nanoparticle agglomeration. In the methodology offered here, it is supposed that CNTs have an affinity to agglomerate when introduced into a polymer matrix, forming spherical-shaped inclusions that can be initiated both in agglomerated areas and dispersed throughout the matrix. This unique nature is described in detail in (Shi et al. 2004). In other words, if W_r represents the overall volume of CNTs in the matrix, it can be represented as the sum of two contributions, as stated by Eq. 4.39.

$$W_r = W_r^{in} + W_r^m \tag{4.39}$$

The given equations express the volume and mass fractions of CNTs and the matrix in a composite material. The volumes of nanofibers inside the inclusions and distributed in the matrix are represented by W_r^{in} and W_r^m, respectively. The total volume of *a* illustrative element, W, is the sum of the nanofiber volume and the matrix volume, as given by Eq. 4.40.

$$W = W_r + W_m \tag{4.40}$$

The mass fraction of CNTs, w_r, and the matrix, w_m, can also be defined in Eq. 4.41.

$$w_r = \frac{M_r}{M_r + M_m} \quad \text{and} \quad w_m = \frac{M_m}{M_r + M_m} \tag{4.41}$$

Similarly, the volume fraction of CNTs, V_r, and the matrix, V_m, can be expressed in terms of their corresponding mass fractions as Eq. 4.42.

$$V_r = \frac{W_r}{W} \quad \text{and} \quad V_m = \frac{W_m}{V} \tag{4.42}$$

The reinforcing phase V_r is the quantity of the volume fraction of CNTs inside the inclusions V_r^{in} and the scattered nanoparticles in the matrix, V_r^m as Eq. 4.43.

$$V_r = V_r^m + V_r^{in} \tag{4.43}$$

The volume fraction of CNTs can be defined as a function of the corresponding mass fraction, w_r, and the density of CNTs and the matrix, denoted by ρ_r and ρ_m, respectively, as Eq. 4.44:

$$V_r = \left(\frac{\rho_r}{w_r \rho_m} - \frac{\rho_r}{\rho_m} + 1 \right)^{-1} \tag{4.44}$$

The agglomeration of particles in the composite material is governed by two parameters as designated in the work by Shi et al. (2004) as Eq. 4.45.

$$\mu = \frac{W_{in}}{W} \text{ and } \eta = \frac{W_r^{in}}{W_r} \tag{4.45}$$

The parameters η and μ are used to characterize the agglomeration of CNTs within the composite material. The value of μ determines the spherical inclusion width W_{in} in relation to the entire volume W, while η quantifies the volume of CNTs inside the inclusions W_r^{in} relative to the total volume of CNTs W_r.

There are three possible cases of agglomeration ratio depending on the values of η and μ. If μ is less than η and less than 1, then the agglomeration is limited, and the nanofibers are both encompassed in the inclusions and dispersed in the matrix. This scenario amplifies the heterogeneity of CNTs, particularly when η increases. On the other hand, when μ is set to be equal to η or 1, the volume of CNTs is entirely focused within the inclusions, and there is no agglomeration between CNTs in the matrix. Finally, complete agglomeration is defined when μ is less than η, and η is equal to 1, where all the CNTs are located within the spherical inclusions. The relationships between the different parameters can be obtained by combining Eqs. 4.46 and 4.47.

$$\frac{W_r^{in}}{W_{in}} = \frac{V_r \eta}{\mu} \tag{4.46}$$

$$\frac{W_r^m}{W - W_{in}} = \frac{V_r (1 - \eta)}{1 - \mu} \tag{4.47}$$

Suppose that CNTs are randomly distributed within the matrix. In this case, the overall material properties of the composite material with spherical inclusions are determined using the Eshelby–Mori–Tanaka approach. This approach involves calculating the bulk modulus of the spherical inclusions K_{in}^* given by Eq. 4.48.

$$K_{in}^* = K_m + \frac{V_r \eta (\eta_r - 2 G_m \beta_r)}{2 (\mu - V_r \eta + V_r \eta \beta_r)} \tag{4.48}$$

while the respective shear modulus G_{in}^* is given by Eq. 4.49.

$$G_{in}^* = G_m + \frac{V_r \eta (\eta_r - 2 G_m \beta_r)}{2 (\mu - V_r \eta + V_r \eta \beta_r)} \tag{4.49}$$

K_m and G_m denote the bulk and the shear moduli of the single isotropic matrix, which can be evaluated as Eq. 4.50.

$$K_m = \frac{E_m}{3(1-2v_m)} \quad \text{and} \quad G_m = \frac{E_m}{2(1+v_m)} \tag{4.50}$$

Bulk modulus K_{out}^* of the matrix scattered with CNT and shear modulus G_{out}^* can be stated as Eqs. 4.51 and 4.52.

$$K_{out}^* = K_m + \frac{V_r(1-\eta)(\delta_r - 3K_m\alpha_r)}{3\left[1-\mu-V_r(1-\eta)+V_r(1-\eta)\alpha_r\right]} \tag{4.51}$$

$$G_{out}^* = G_m + \frac{V_r(1-\eta)(\eta_r - 2G_m\beta_r)}{2\left[1-\mu-V_r(1-\eta)+V_r(1-\eta)\beta_r\right]} \tag{4.52}$$

The values α_r, β_r, δ_r, η_r are estimated by Eqs. 4.53–4.56.

$$\alpha_r = \frac{3(K_m + G_m) + k_r + l_r}{3(G_m + k_r)} \tag{4.53}$$

$$\beta_r = \frac{1}{5}\left[\frac{4G_m + 2k_r + l_r}{3(G_m + k_r)} + \frac{4G_m}{G_m + p_r} + \frac{2(G_m(3K_m + G_m) + G_m(3K_m + 7G_m))}{G_m(3K_m + G_m) + m_r(3K_m + 7G_m)}\right] \tag{4.54}$$

$$\delta_r = \frac{1}{3}\left[\eta_r + 2l_r + \frac{(2k_r + l_r)(3K_m + G_m - l_r)}{G_m + k_r}\right] \quad \text{and}$$

$$\eta_r = \frac{1}{5}\left[\frac{2}{3}(\eta_r - l_r) + \frac{8G_m p_r}{G_m + p_r} + \frac{(2k_r - l_r)(2G_m + l_r)}{2(G_m + k_r)} + \frac{8m_r G_m(3K_m + 4G_m)}{3K_m(m_r + G_m) + G_m(7m_r + G_m)}\right] \tag{4.55}$$

Total bulk and shear modulus of the CNT reinforced matrix are estimated by Eqs. 4.56 and 4.57.

$$K_m^* = K_{out}^*\left[1 + \frac{\mu\left(\dfrac{K_{in}^*}{K_{out}^*} - 1\right)}{1 + (1-\mu)\left(\dfrac{K_{in}^*}{K_{out}^*} - 1\right)\left(\dfrac{1 + v_{out}^*}{3 - 3v_{out}^*}\right)}\right] \tag{4.56}$$

$$G_m^* = G_{out}^*\left[1 + \frac{\mu\left(\dfrac{G_{in}^*}{G_{out}^*} - 1\right)}{1 + (1-\mu)\left(\dfrac{G_{in}^*}{G_{out}^*} - 1\right)\left(\dfrac{8 - 10v_{out}^*}{15 - 15v_{out}^*}\right)}\right] \tag{4.57}$$

Here, Poisson's ratio, $v_{out}^* = \left(3K_{out}^* - 2G_{out}^*\right)\Big/\left(6K_{out}^* - 2G_{out}^*\right)$

With the consideration of modified matrix to be isotropic, elastic modulus (E_m^*), Poisson's ratio (v_m^*) and density (ρ_m^*) are estimated by Eqs. 4.58–4.60 as

$$E_m^* = \frac{9K_m^* \, G_m^*}{3K_m^* + G_m^*} \tag{4.58}$$

$$v_m^* = \frac{3K - 2G}{6K + 2G} \tag{4.59}$$

$$\rho_m^* = \left(\rho_r - \rho_m\right)V_r + V_m \tag{4.60}$$

It is essential to focus on the fact that the volume fraction of CNTs V_r can also be varied gradually to create functionally graded CNT (FGCNT)-reinforced composites, where each mechanical property depends on the direction of the gradual variation. However, for the purpose of this study, a continuous distribution of CNT volume fraction is presumed as a hypothesis. With this hybrid isotropic matrix established, the next step involves combining it with strengthening fibers to create the preferred three-phase multiscale composite.

4.2.3 HOMOGENIZATION TECHNIQUES (HTs)

The HT refers to standardizing the mechanical properties of fiber-reinforced modified matrix composites. Modified matrix refers to modifying the matrix properties by incorporating CNT into them. Basically, the elastic property of a matrix is in certain MPa and the CNT in TPa; when these are combined to form a homogeneous modified matrix, the overall property of the newly formed matrix will be greatly improved. To evaluate these properties, one needs to consider the elastic properties of the hybrid matrix (E_m^*, v_m^*, and ρ_m^*) as well as those required to characterize the reinforcing fibers. Typically, reinforcing fibers are assumed to be transversely isotropic, which requires five independent elastic constants to fully characterize. In terms of engineering material parameters, this includes the Young's moduli ($E_1^f = E_2^f = E_3^f$), the shear moduli ($G_{12}^f = G_{13}^f = G_{23}^f$), and the Poisson's ratios ($v_{12}^f = v_{13}^f = v_{23}^f$).

The fiber volume fraction (V_f) and its corresponding density (ρ_f) are also required for mechanical characterization. For anisotropic fibers such as Carbon and Kevlar, more complex calculations are required. The V_f and ρ_f are also required for mechanical characterization, and V_f can be conveyed as a function of the corresponding mass fraction of the fibers (w_f), calculated by Eq. 4.61.

$$V_f = \left(\frac{\rho_f}{w_f \rho_m^*} - \frac{\rho_f}{\rho_m^*} + 1\right)^{-1} \tag{4.61}$$

Once V_f is determined, the volume fraction of the hybrid matrix can be calculated by $V_m^* = 1 - V_f^*$. The mechanical parameters of a multiphase, multiscale composite, including the Young's moduli, shear moduli, and Poisson's ratios, can be evaluated through the subsequent equations. It must be noted that the following criteria should be verified with Eqs. 4.62 and 4.63.

$$\Delta = 1 - v_{12}v_{21} - v_{23}v_{32} - v_{13}v_{31} - 2v_{21}v_{32}v_{13} \tag{4.62}$$

$$0 < v_{ij} < \sqrt{\frac{E_i}{E_j}} \text{ for } i \neq j \tag{4.63}$$

When analyzing a laminated structural composite, it is essential to note that each engineering constant must be attributed to the specific layer being considered. To estimate the density of the structure, the rule of mixture can be utilized. At this point, it is necessary to recall some basic assumptions outlined in Chamis and Sendeckyj (1968). From a macroscopic perspective, the composite is assumed to be linearly elastic and transversely isotropic, and the occurrence of enduring stresses in a stress-free condition is not permitted. Additionally, It is assumed that the matrix and the reinforcing fibers are linearly elastic, consistent, and totally devoid of cavities. The interfaces between the matrix and fibers are perfectly united, and it is assumed that the fibers are regularly aligned and placed within the matrix.

4.2.4 CHAMIS APPROACH

The following relationships enable the determination of the elastic characteristics of a single directionally oriented lamina composed of anisotropic fibers embedded in an isotropic matrix, as shown in Eqs. 4.64–4.69.

$$E_1 = E_1^f V_f + E_m^* V_m^* \tag{4.64}$$

$$E_2 = E_3 = \frac{E_m^*}{1 - V_f\left(1 - E_m^*/E_2^f\right)} \tag{4.65}$$

$$G_{12} = G_{13} = \frac{G_m^*}{1 - V_f\left(1 - G_m^*/G_{12}^f\right)} \tag{4.66}$$

$$G_{23} = \frac{G_m^*}{1 - V_f\left(1 - G_m^*/G_{23}^f\right)} \tag{4.67}$$

$$v_{12} = v_{13} = v_{12}^f V_f + v_m^* V_m^* \tag{4.68}$$

$$v_{23} = \frac{E_2}{2G_{23}} - 1 \tag{4.69}$$

The equations mentioned in this text can be found in Chamis' works (Chamis 1969, 1983). A transversely isotropic medium supposition leads to the similarity of properties in the second and third directions, which necessitates five independent constants for characterizing the composite's mechanical behavior. This homogenization method is widely used due to its simplicity and is referred to as the "rule of mixture" in many contemporary literary works. However, it should be noted that all methods based on the mechanics of materials can generally be regarded as the rule of mixture. Chamis and Sendeckyj (1968) provide additional information about the mechanics of materials. Refinements to expressions can be made to improve the accuracy of mechanical property estimations. For instance, the relation presented below can be utilized to determine the values of $E_2 = E_3$, as indicated in Eq. 4.70.

$$\frac{1}{E_2} = \frac{V_f}{E_2^f} + \frac{V_m^*}{E_m^*} - V_f V_m^* \frac{\left(v_{12}^f\right)^2 \left(\frac{E_m^*}{E_2^f}\right) + \left(v_m^*\right)^2 \left(\frac{E_2^f}{E_m^*}\right) - 2v_{12}^f v_m^*}{V_f E_2^f + V_m^* E_m^*} \qquad (4.70)$$

4.2.5 HAHN APPROACH

The Hahn (1980) approach is a homogenization method that can be employed to determine the mechanical characteristics of a fibrous composite consisting of randomly arranged fibers with circular cross-sections perpendicular to the direction of oriented fibers. This results in a macroscopic, transversely isotropic composite that necessitates five independent elastic constants for characterization. The parameters in Eqs. 4.71–4.73 are necessary:

$$\Delta_1 = \frac{1 + G_m^* / G_{12}^f}{2} \qquad (4.71)$$

$$\Delta_2 = \frac{3 - 4v_m^* + G_m^* / G_{23}^f}{4\left(1 - v_m^*\right)} \qquad (4.72)$$

$$\Delta_k = \frac{1 + G_m^* / K_f}{2\left(1 - v_m^*\right)} \qquad (4.73)$$

here, bulk modulus of fiber reinforcement, $K_f = E_2^f / 3\left(1 - 2v_{23}^f\right)$.

Plane strain bulk modulus of the composite (K_T) with consideration of bulk modulus of hybrid matrix (K_m^*) is given in Eq. 4.73 as

$$K_T = \frac{V_f + \Delta_K V_m^*}{\dfrac{V_f}{K_f} + \dfrac{\Delta_K V_m^*}{K_m^*}} \qquad (4.73)$$

4.2.6 Hashin–Rosen Approach

In the publication by Hashin and Rosen (1964), a homogenization method based on a variational approach is presented, which is a straightforward technique for determining the mechanical material properties of a fiber-reinforced composite (fiber oriented in the same direction) with isotropic characteristics for both the fibers and the matrix. As previously mentioned, this assumption is appropriate for E-glass fibers. However, it is not suitable for Carbon or Kevlar fibers, which are transversely isotropic. The engineering constants for this method are stated in Eqs. 4.74–4.79 as:

$$E_1 = E_f V_f + E_m^* V_m^* + \frac{4 V_f V_m^* \left(v_f - v_m^*\right)^2}{\dfrac{V_m^*}{\overline{k}_f} + \dfrac{V_f}{\overline{k}_m^*} + \dfrac{1}{G_m^*}} \tag{4.74}$$

$$E_2 = E_3 = \frac{4 \overline{k}_t\, G_t}{\overline{k}_t + G_t \left(1 + \dfrac{4 \overline{k}_t \left(v_{12}\right)^2}{E_{11}} \right)} \tag{4.75}$$

$$G_{12} = G_{13} = G_m^* \frac{V_m^* G_m^* + \left(1 + V_f\right) G_f}{\left(1 - V_f\right) G_m^* + V_m^* G_f} \tag{4.76}$$

$$G_{23} = \frac{E_2}{2\left(1 + v_{23}\right)} \tag{4.77}$$

$$v_{12} = v_{13} = v_f V_f + v_m^* V_m^* + \frac{V_f V_m^* \left(v_f - v_m^*\right)/\left(\dfrac{1}{\overline{k}_m^*} + \dfrac{1}{\overline{k}_f}\right)}{\dfrac{V_m^*}{\overline{k}_f} + \dfrac{V_f}{\overline{k}_m^*} + \dfrac{1}{G_m^*}} \tag{4.78}$$

$$v_{23} = \frac{E_2}{2G_t} - 1 \tag{4.79}$$

The corresponding values of different parameters are given in Eqs. 4.80–4.87 as

$$\overline{k}_f = \frac{E_f}{2\left(1 - v_f - \left(v_f\right)^2\right)} \tag{4.80}$$

$$\overline{k}_m^* = \frac{E_m^*}{2\left(1 - v_m^* - \left(v_m^*\right)^2\right)} \tag{4.81}$$

$$\overline{k}_t = \frac{\overline{k}_m^* \overline{k}_f + \left(V_f \overline{k}_f + V_m^* \overline{k}_m^*\right) G_m^*}{V_m^* \overline{k}_f + V_f \overline{k}_m^* + G_m^*} \tag{4.82}$$

$$G_t = G_m^* \frac{\left(\alpha + \beta_m^* V_f\right)\left(1 + \xi\left(V_f\right)^3\right) - 3V_f\left(V_m^* \beta_m^*\right)^2}{\left(\alpha - V_f\right)\left(1 + \xi\left(V_f\right)^3\right) - 3V_f\left(V_m^* \beta_m^*\right)^2} \tag{4.83}$$

$$\alpha = \frac{G_f / G_m^* + \beta_m^*}{G_f / G_m^* - 1} \tag{4.84}$$

$$\beta_m^* = \frac{1}{3 - 4v_m^*} \tag{4.85}$$

$$\beta = \frac{1}{3 - v_f} \tag{4.86}$$

$$\xi = \frac{\beta_m^* - \beta_f G_f / G_m^*}{1 + \beta_f G_f / G_m^*} \tag{4.87}$$

To ensure all relevant information is included, it is worth noting that $\bar{k}_f$ and $\bar{k}_m^*$ represent the bulk moduli for the fibers and hybrid matrix under plane strain conditions, respectively.

4.2.7 HALPIN–TSAI APPROACH

A technique known as homogenization, named after the studies by Halpin (1969) and Tsai (1964, 1965), is utilized to determine the mechanical material properties of the multiphase composite. This method employs Hill's elastic moduli and a semi-empirical approach. Specifically, when considering the reinforcing fibers, Hill's flexible moduli are expressed in Eqs. 4.88–4.92:

$$k_f = \frac{E_2^f}{2\left(1 - v_{23}^f - 2v_{21}^f v_{12}^f\right)} \tag{4.88}$$

$$l_f = \frac{v_{12}^f E_2^f}{\left(1 - v_{23}^f - 2v_{21}^f v_{12}^f\right)} = 2v_{12}^f k_f \tag{4.89}$$

$$m_f = \frac{E_2^f}{2\left(1 + v_{23}^f\right)} \tag{4.90}$$

$$n_f = \frac{E_1^f\left(1 - v_{23}^f\right)}{\left(1 - v_{23}^f - 2v_{21}^f v_{12}^f\right)} = 2\left(1 - v_{23}^f\right)\frac{E_1^f}{E_2^f} k_f \tag{4.91}$$

$$p_f = G_{12}^f \tag{4.92}$$

where k_f, l_f, m_f, n_f, p_f represent the fiber parameters in terms of Hill's moduli. On the other side, if the isotropic hybrid matrix is considered, then revised as Eqs. 4.92–4.96.

$$k_m^* = \frac{E_m^*}{2\left(1+v_m^*\right)\left(1-2v_m^*\right)} \tag{4.92}$$

$$l_m^* = \frac{v_m^* E_m^*}{2\left(1+v_m^*\right)\left(1-2v_m^*\right)} = 2v_m^* k_m^* \tag{4.93}$$

$$m_m^* = \frac{E_m^*}{2\left(1+v_m^*\right)} \tag{4.94}$$

$$n_m^* = \frac{E_m^*\left(1-v_m^*\right)}{\left(1+v_m^*\right)\left(1-2v_m^*\right)} \tag{4.95}$$

$$p_m^* = G_m^* = \left(1-2v_m^*\right)k_m^* \tag{4.96}$$

where k_f, l_f, m_f, n_f, p_f signify the properties of the hybrid matrix in terms of Hill's moduli. The cumulative material properties of the multiphase composite is given in Eqs. 4.97–4.101:

$$k = \frac{k_m^*\left(k_f + m_m^*\right)V_m^* + k_f\left(k_m^* + m_m^*\right)V_f}{\left(k_m^* + m_m^*\right)V_m^* + \left(k_m^* + m_m^*\right)V_f} \tag{4.97}$$

$$l = V_f l_f + V_m^* l_m^* + \frac{l_f - l_m^*}{k_f - k_m^*}\left(k - V_f k_f - V_m^* k_m^*\right) \tag{4.98}$$

$$m = m_m^* \frac{2V_f m_f\left(k_m^* + m_m^*\right) + 2V_m^* m_f\ m_m^* + V_m^* k_m^*\left(m_f + m_m^*\right)}{2V_f m_m^*\left(k_m^* + m_m^*\right) + 2V_m^* m_f\ m_m^* + V_m^* k_m^*\left(m_f + m_m^*\right)} \tag{4.99}$$

$$n = V_f n_f + V_m^* n_m^* + \left(\frac{l_f - l_m^*}{k_f - k_m^*}\right)^2\left(k - V_f k_f - V_m^* k_m^*\right) \tag{4.100}$$

$$p = \frac{\left(p_f + p_m^*\right)V_m^* p_m^* + 2\ p_f p_m^* V_f}{\left(p_f + p_m^*\right)V_m^* + 2p_m^* V_f} \tag{4.101}$$

Thus, engineering constants of the composites are stated in Eqs. 4.102-4.107:

$$E_1 = n - \frac{l^2}{k} \tag{4.102}$$

$$E_2 = E_3 = \frac{4m\left(kn - l^2\right)}{kn - l^2 + mn} \tag{4.103}$$

$$v_{12} = v_{13} = \frac{1}{2k} \tag{4.104}$$

$$v_{23} = \frac{n\left(k - m\right) - l^2}{n\left(k + m\right) - l^2} \tag{4.105}$$

$$G_{12} = G_{13} = p \tag{4.106}$$

$$G_{23} = m \tag{4.107}$$

4.3   BUCKLING BEHAVIOR OF SMPCs

When subjected to compressive loads, traditional fiber-reinforced composites can experience a range of failure modes, such as shear or tensile failure, fiber disruption, and fiber brittle rupture (Tiwari and Shaikh, 2021a, 2021b). Comparatively, general resin composites exhibit higher stiffness and a narrower range of buckling deformation, whereas fiber-reinforced SMPCs demonstrate a hyperelastic nature at high temperatures $(T > T_g)$, and their fiber buckling distortion capacity increases.

During the SMC, SMPCs are inevitably subjected to bending distortion with high curvatures at elevated temperatures. Under these conditions, the compression zone of SMPC undergoes microscopic buckling, which impacts the macroscopic stiffness of the structure.

4.3.1   PREPARATION OF STRAIN IN SMPC PLATES

The study focused on uniformly orienting carbon fibers in an SMP matrix, which is a polymer based on acrylate, along the x-axis. During the bending test, it was assumed that the connection between the reinforcement and matrix was flawless, and there was no consideration given to the possibility of the SMPC structure being damaged or failing due to the force that was applied in the y direction. The examination was carried out using SMPC plates that had the following dimensions: length l, width w, and thickness t. In the beginning, there was no microbuckling of the fibers, and the neutral surface corresponded to the mid-surface layer parallel to the plate surface. Furthermore, the dimensions of the neutral surface did not change in any way. However, once the amount of bending exceeded a certain threshold, the fibers on the compression side began to buckle, which caused the neutral surface to shift toward the tensile side in order to keep the deformed plate's equilibrium. This phase formed three zones: one with buckled fibers on the compression side $(-t/2, z_b)$, one with unbuckled fibers on the compression side (z_b, z_n), and one with unbuckled fibers on the tension side $(z_n, t/2)$, as illustrated in Figure 4.9. The neutral strain seeming and the acute buckling surface were identified by their respective coordinates, z_n and z_b. The microbuckling of the reinforced fiber is depicted in Figure 4.10. (Zhang, Dui, and Liang 2018).

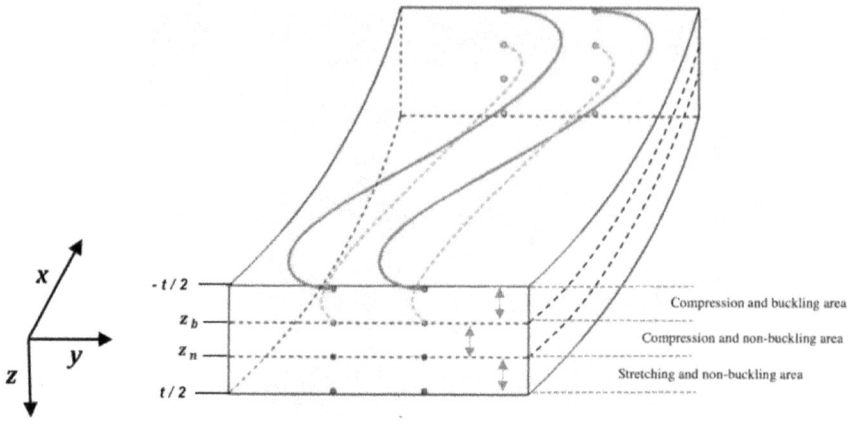

FIGURE 4.9 The buckling fibers in the SMPC plate (Tiwari and Shaikh 2021c).

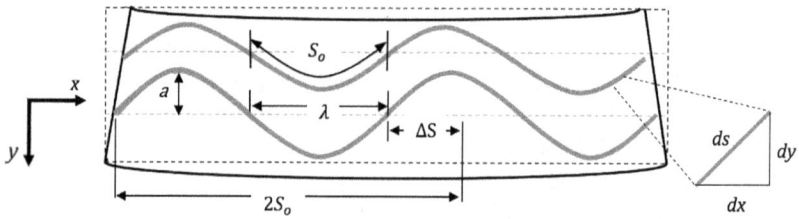

FIGURE 4.10 Microbuckling of the fiber in the SMPC underneath large strain distortion (Tiwari and Shaikh 2021c).

When the SMPC plate is bent, the fiber experiences microbuckling within the plane, causing the fiber to shift harmonically along the y-axis (i.e., the width of the plate). This sinusoidal path taken by the micro-sideway deflection fiber is referred to as Eq. 4.108:

$$y(x) = a\,\sin\left(\frac{\pi x}{\lambda}\right) \tag{4.108}$$

The magnitude of the sinusoidal wave of the suddenly distorted strands/fibers is denoted by 'a', while 'λ' represents the equivalent half wavelength. The fiber is assumed to be incompressible, resulting in equal overall length before and after buckling. Eq. 4.109 expresses the arc length by substituting the differential arc length from the Pythagorean theorem.

$$S_o = \int_0^{S_o} ds = \int_0^{\lambda} \sqrt{dx^2 + dy^2}\,dx \tag{4.109}$$

The engineering strain (ε_{xx}) is calculated by Eq. 4.110 as

$$\varepsilon_{xx} = \frac{\Delta S}{2S_o} = 1 - \frac{\lambda}{S_o} \qquad (4.110)$$

Kirchhoff's theory states that the strain experienced by a bent SMPC plate is proportional to a function of the neutral strain zone face (z_n). Eq. 4.111 is linear throughout the thickness of the plate.

$$\varepsilon_{xx} = k(z - z_n) \qquad (4.111)$$

The value of the mid-plane bending arch (k) in meters inverse and the strain in the x direction (ε_{xx}) are used to calculate the amplitude (a) of the sinusoidal deflected fiber for positive values of $(z_n - z)$. Francis has presented Eq. 4.112 as a function that relates the amplitude (a) to thickness (Francis et al. 2012).

$$a(z) = \frac{2\lambda}{\pi} \sqrt{k(z_n - z)} \qquad (4.112)$$

The amount of the dislocation of the microbuckled fiber in the y direction may be calculated using Eq. 4.113:

$$y(x,z) = \frac{2\lambda}{\pi} \sqrt{k(z_n - z)} \, \sin\left(\frac{\pi x}{\lambda}\right) \qquad (4.113)$$

Elastic deformation in the matrix is negligible compared to the equivalent shear deformation when the SMPC panel is in the buckling region. Figures 4.11 and 4.12 depict the shear strain in the matrix as γ_{xy} and γ_{yz}, respectively, caused by the fiber's z-axis displacement in the xy and yz planes. There is no longer any indefinite shear component in the xz region. And because fiber stiffness exceeds matrix stiffness by a

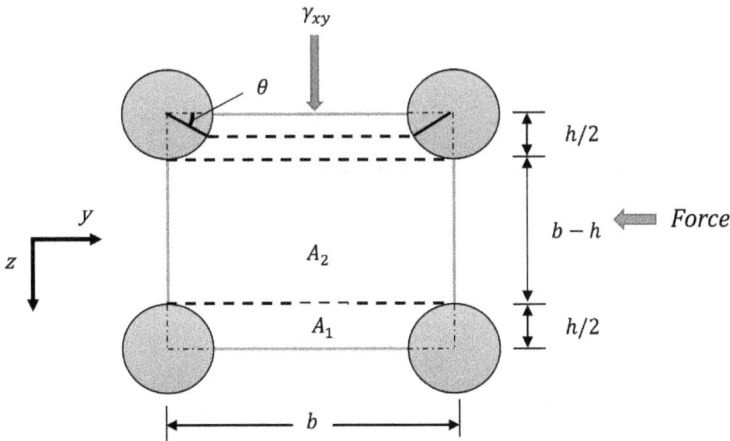

FIGURE 4.11   Alteration of the unit cell produced by strain in the xy plane (γ_{xy}) (Tiwari and Shaikh 2021c).

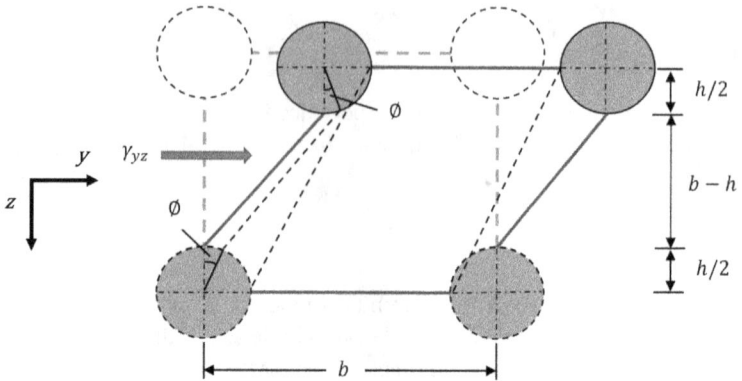

FIGURE 4.12 Alteration of the unit cell caused by strain in yz plane (γ_{yz}) (Tiwari and Shaikh 2021c).

wide margin, matrix displacement during buckling may be neglected. In the unbuckling region, the x-axis deformation is attributable to stretching, whereas the y- and z-axis components are also zero.

When defining the values of γ_{xy} and γ_{yz}, it is assumed that the fiber is a three-dimensional object, and its dimensions are taken into account. The term "shear strain along the xy plane" (γ_{xy}) refers to a specific type of deformation that occurs within the fiber as stated in Eq. 4.114.

$$\gamma_{xy} = \frac{\partial v}{\partial x} + \frac{\partial u}{\partial y} \tag{4.114}$$

Filed vectors along x-and y-directions are shown by u and v, respectively.

The matrix distortion around each fiber is not homogeneous because of the significance of its fiber size. Thus, established on the figure shown in Figure 4.13, γ_{xy} is given by Eq. 4.115.

$$\gamma_{xy} = \frac{\partial y}{\partial x} + \frac{\Delta u}{\Delta y} \tag{4.115}$$

The displacement values, Δu and Δy, are given by Eqs. 4.116 and 4.117.

$$\Delta u = \frac{h\cos\theta}{2}\frac{\partial y}{\partial x} \tag{4.116}$$

$$\Delta y = \frac{b - h\cos\theta}{2} \tag{4.117}$$

For optical fibers, b is the distance in millimeters between the centers of two consecutive fibers, and h is the fiber's diameter. Additionally, the angular measurement ranges from 0° to 90°, with 0° being measured along the horizontal axis (x-axis).

For a given fiber diameter h, the length of its horizontal projection parallel to the y-axis is given by $h\cos\theta$ (as shown in Figure 4.13).

$$b = \sqrt{\frac{\pi h^2}{4V_f}} \tag{4.118}$$

where percentage of the SMPC plate that is made up of fiber reinforcement, denoted as V_f.

For the plane angle $0°$ and $90°$ parallel to y-axis, the shear strain stated as Eq. 4.119

$$\gamma_{xy} = \frac{h\cos\theta}{b-h\cos\theta}\frac{\partial y}{\partial x} + \frac{\partial y}{\partial x} = \frac{b}{b-h\cos\theta}\frac{\partial y}{\partial x} \tag{4.119}$$

The lowest and highest values of γ_{xy} are expressed as Eqs. 4.120 and 4.121

$$\gamma_{xy}^{max} = \max(\gamma_{xy}) = \frac{b}{b-h}\frac{\partial y}{\partial x} \tag{4.120}$$

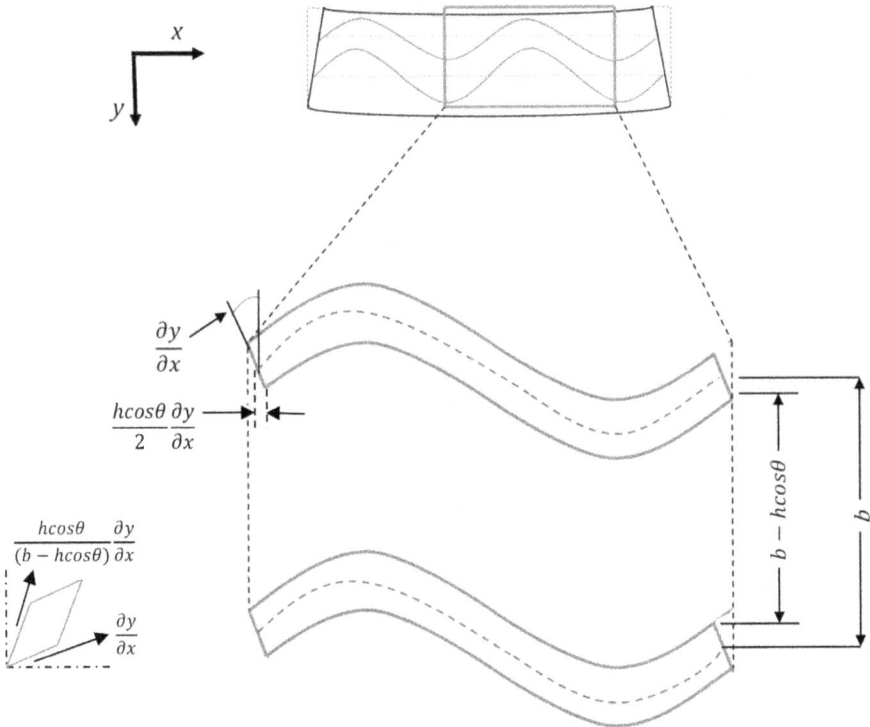

FIGURE 4.13 Deformation due to strain in the xy plane (γ_{xy}) on a magnified section (Tiwari and Shaikh 2021c).

$$\gamma_{xy}^{min} = \min(\gamma_{xy}) = \frac{\partial y}{\partial x} \tag{4.121}$$

Then the γ of the SMP polymer matrix in the yz plane is given as Eq. 4.122,

$$\gamma_{yz} = \frac{\partial v}{\partial z} + \frac{\partial w}{\partial y} \tag{4.122}$$

Figure 4.12 illustrates that during the microbuckling of fibers, the amplitude of sinusoidal trajectories is significant at various positions in the z-direction. As a result, it can be stated that γ_{yz} is primarily influenced by Eqs. 4.123–4.125.

$$\gamma_{yz} = \frac{\Delta y}{\Delta z} \tag{4.123}$$

$$\Delta y = \Delta a(z)\sin\left(\frac{\pi x}{\lambda}\right) \tag{4.124}$$

$$\Delta z = b - h\cos\varnothing \tag{4.125}$$

The angular amount of the movement of the matrix underneath strain is denoted by $\varnothing$ and ranges from $0°$ to $90°$. The span of the transverse projection of fiber diameter h, which is parallel to the z-axis, is represented by $h\cos\varnothing$. Based on this information, the value of Δa can be expressed as Eq. 4.126,

$$\Delta a(z) = a(z) - \Delta a(z - b) \tag{4.126}$$

Thus, Eq. 4.127 for γ_{yz} for $0° \leq \varnothing \leq 90°$ is

$$\gamma_{yz} = \frac{\Delta a(z)}{b - h\cos\varnothing}\sin\left(\frac{\pi x}{\lambda}\right) \tag{4.127}$$

The highest and lowest value of γ_{yz} at $0° \leq \varnothing \leq 90°$ are estimated by Eqs. 4.128 and 4.129.

$$\gamma_{yz}^{min} = \min(\gamma_{yz}) = \frac{\Delta a(z)}{b}\sin\left(\frac{\pi x}{\lambda}\right) \tag{4.128}$$

$$\gamma_{yz}^{max} = \max(\gamma_{yz}) = \frac{\Delta a(z)}{b - h}\sin\left(\frac{\pi x}{\lambda}\right) \tag{4.129}$$

The amount of the strain in SMPC along the x-axis for neutral plane bending curvature (k) is given by Eq. 4.130,

$$\varepsilon_{xx} = k(z - z_n) \tag{4.130}$$

4.3.2 MATHEMATICAL INVESTIGATION OF THE STRAIN ENERGY IN THE SMPC PANEL

When analyzing the microbuckling of an SMPC plate, the cumulative strain energy (U) can be divided into two constituents: temperature-dependent and temperature-independent factors. The temperature-dependent factor consists of three parts, namely the strain energy of the matrix along the yz and xy plane (U_{yz}, U_{xy}), and the elongated strain energy of the SMPC (U_{str}). On the other hand, the strain energy of the fiber (U_f) was the only temperature self-governing module that was taken into consideration during the analysis of microbuckling. This is stated as Eq. 4.131

$$U = U_{yz} + U_{xy} + U_{str} + U_f \tag{4.131}$$

4.3.3 SHEAR STRAIN ENERGY IN THE XY PLANE

Figure 4.3 shows an area A_1 on the balanced side of the single entity cell where the shear strain energy of the SMP matrix alongside the xy plane can be calculated. The expression for this energy is given by Eq. 4.132.

$$u_{xy} = \tau_{xy} \gamma_{xy} A_1 \tag{4.132}$$

where $A_1 = \int_0^{\pi/2} \left[\dfrac{h}{2}(b - h\cos\theta)\cos\theta \right] d\theta.$

Equivalent U in the remaining area (A_2) of the unit cell is assumed in Eq. 4.133 as

$$\tilde{u}_{xy} = \frac{1}{2}\tau_{xy}^{min}\,\gamma_{xy}^{min} A_2 \tag{4.133}$$

where, $A_2 = b(b - h).$

$\tilde{U}_{xy}$ of the SMP matrix in the exterior feature parallel to the xy plane is given by Eq. 4.134.

$$\tilde{U}_{xy} = \frac{u_{xy} + \tilde{u}_{xy}}{b^2} \tag{4.134}$$

The entire shear strain of the SMP along the whole buckling area of the composite is assumed by Eq. 4.135.

$$U_{xy} = \iiint \tilde{U}_{xy}\,dV \tag{4.135}$$

This is simplified as Eq. 4.136.

$$U_{xy}(T, z_n, z_b) = G_m(T)\,\mathrm{kwlf}\big(V_f\big)g\big(z_n, z_b\big) \tag{4.136}$$

The above expression $G_m(T)$ gives the heat-operated shear modulus of the SMP matrix, k, which denotes the mid-plane bending curvature, and w and l, which respectively state the width and length of the SMPC plate. Additionally, the equation takes into account the distances z_n and z_b, which represent the distance to the neutral strain surface from the upper surface and the distance from the upper surface to the critical buckling surface, respectively. The calculation also considers the microbuckling shape-dependent on factors $f(V_f)$ and $g(z_n, z_b)$, presented by Eqs. 4.137 and 4.138.

$$f(V_f) = 1 - \frac{\pi}{2} - \sqrt{\frac{4V_f}{\pi}} + \frac{2\sqrt{\pi}}{\sqrt{\pi - 4V_f}} \arctan\left(\frac{\sqrt{\pi - 4V_f}}{\sqrt{\pi} - \sqrt{4V_f}}\right) \qquad (4.137)$$

$$g(z_n, z_b) = \left(z_n z_b + \frac{z_n t}{2} - \frac{z_b^2}{2} + \frac{t^2}{2}\right) \qquad (4.138)$$

4.3.4 SHEAR STRAIN ENERGY IN THE YZ PLANE

Using a method similar to the one utilized for estimating U_{xy}, the shear strain energy of the SMP matrix in the yz plane U_{yz} in the buckling zone can be determined. The expression for this temperature-regulated energy is provided by Eq. 4.139.

$$U_{yz}(T, \lambda, z_n, z_b) = \frac{G_m(T) k w l \lambda^2}{\pi^2} f(V_f) f(z_n, z_b) \qquad (4.139)$$

The calculation of the heat-operated shear strain energy in the yz plane (U_{yz}) for the buckling zone involves the half wavelength (λ) of the microbuckled fiber in the xy plane. The expression also includes a geometric factor $f(z_n, z_b)$, which takes into account the distance from the upper surface to the neutral strain surface (z_n) and the distance from the upper surface to the critical buckling surface (z_b). It should be noted that the approximation is that the fiber spacing (b) is considerably smaller than the total thickness (t).

$$f(z_n, z_b) = \frac{1}{4b^2}\left[-(2z_n - 2z_b + b)^2 + 2(2z_n - 2z_b + b)\sqrt{(z_n - z_b)(z_n - z_b + b)} \right.$$

$$\left. + b^2 \ln\left(\frac{2(t + 2z_n + b)}{2\sqrt{(z_n - z_b)(z_n - z_b + b)}(2z_n - 2z_b + b)}\right)\right]$$

4.3.5 STRETCH STRAIN ENERGY OF THE SMPC

Stretching the region below z_n in the glass transition zone of the investigated SMP results in a strain energy in the SMPC plate that is represented as Eq. 4.141.

$$U_{str} = \frac{1}{2}\iiint \sigma_{xx}\varepsilon_{xx}\, dV \qquad (4.141)$$

After integration, this is represented by Eq. 4.142.

$$U_{str}\left(T,z_n,z_b\right)=\frac{E_T(T)wl}{24}k^2H\left(z_n,z_b\right) \tag{4.142}$$

where $H\left(z_n,z_b\right)=\left(t^3-6t^2z_n+12tz_n^2+24z_nz_b^2-24z_n^2z_b-8z_b^3\right)$.

4.3.6 STRAIN ENERGY OF THE FIBERS

The strain energy resulting from the stretching of the SMPC plate in the region below z_n, which lies within the glass transition region of the specific SMP being studied, can be calculated. The expression for this energy is provided by Eq. 4.143.

$$U_{f,s}=\frac{E_fI_f}{2}\int_0^l\left(\frac{\partial^2 y}{\partial^2 x}\right)^2 dx \tag{4.144}$$

where E_f = longitudinal modulus and I_f = fibre inertia

The strain energy density of the buckled fiber, based on the RVE model, is given as Using RVE, the buckeled fiber starin energy shown by Eq. 4.145,

$$\bar{U}_f=\frac{U_{f,s}}{b^2l} \tag{4.145}$$

Based on the integration, the strain energy in the fiber over the entire volume (V) is stated as Eq. 4.146.

$$U_f=\iiint \bar{U}_f\,dV \tag{4.146}$$

The solution of the expression in Eq. 4.146 results in Eq. 4.147.

$$U_f\left(\lambda,z_n,z_b\right)=\frac{4E_fI_fV_f\pi kwl}{\lambda^2h^2}g\left(z_n,z_b\right) \tag{4.147}$$

4.3.7 MICROBUCKLING CONSIDERATIONS OF SMPC

The determination of the unbiased strain surface (z_n), critical buckling surface (z_b), and half wavelength of the fiber (λ) because of the flexure of the SMPC panel can be affected by temperature changes inside the T_g area of the SMP matrix. Below, we provide an estimation of these factors with respect to such temperature variations.

4.3.8 ESTIMATION OF z_n AND z_b

The minimum energy approach was employed to find z_n, U, and z_b. To get closed-form effects, U_{yz} and U_f were neglected. This resulted in the expression Eq. 4.148 for the total strain energy (U_z) along the width at any xy plane.

$$U_z\left(T,z_n,z_b\right)=U_{xy}+U_{str} \tag{4.148}$$

The lowering of Eq. 4.148 to determine z_n and z_b was carried out by assuming that $\partial U_z / \partial z_n = 0$ and $\partial U_z / \partial z_b = 0$. The minimum energy method then provides solutions for z_n and z_b, which are given by Eqs. 4.149 and 4.150:

$$z_n = \frac{t}{2} - \frac{1}{2k} \frac{G_m(T) f(V_f)}{E_t(T)} \left[\sqrt{1 + \frac{4ktE_t(T)}{G_m(T) f(V_f)}} - 1 \right] \quad (4.149)$$

$$z_b = \frac{t}{2} - \frac{1}{2k} \frac{G_m(T) f(V_f)}{E_t(T)} \left[\sqrt{1 + \frac{4ktE_t(T)}{G_m(T) f(V_f)}} + 1 \right] \quad (4.150)$$

The amount of $z_b = z_n - \dfrac{1}{k} \dfrac{G_m(T) f(V_f)}{E_t(T)}$.

4.3.9 EVALUATION OF THE HALF WAVELENGTH FOR FIBERS

To use the minimum energy method, it is necessary to minimize the terms in the aggregate strain energy that involve the half wavelength of the fiber (λ), namely U_{yz} and U_f. This is represented by Eq. 4.151.

$$\frac{\partial U_\lambda}{\partial \lambda} = \frac{\partial}{\partial \lambda} \left[U_f(\lambda, z_n, z_b) + U_{yz}(T, \lambda, z_n, z_b) \right] = 0 \quad (4.151)$$

To simplify the numerical solution, it was assumed that k has an infinite value and $z_n = z_b = t/2$. Therefore, the simplified expression for λ around the T_g region of the SMPC can be given by Eq. 4.152 as:

$$\lambda = \left[\frac{E_f V_f \pi^4 h^2 t^2}{8 G_m(T) f(V_f) \left\{ -1 + \ln\left(\frac{8t}{h} \sqrt{\frac{V_f}{\pi}} \right) \right\}} \right]^{\frac{1}{4}} \quad (4.152)$$

4.4 CASE STUDY: MICROBUCKLING IN SHAPE MEMORY POLYMER COMPOSITES

Gu et al. (2019) proposed the assumption that the SMP matrix is composed of both frozen and energetic parts, whose relative proportions depend on the temperature variation from T_g temperature. Based on this, they presented the temperature-dependent modulus of the matrix as shown in Eq. 4.153:

$$E_m(T) = (E_1 - E_2) \cdot \exp\left(-\left(\frac{T}{T_\beta} \right)^{m_1} \right) + (E_2 - E_3) \cdot \exp\left(-\left(\frac{T}{T_g} \right)^{m_2} \right) + E_3 \cdot \exp\left(-\left(\frac{T}{T_f} \right)^{m_3} \right)$$

$$(4.153)$$

where the storage moduli at T_g, the glass transition temperature, and the flow area after T_g are denoted by E_1, E_2, and E_3, respectively. T_β, T_g, and T_f are the corresponding temperatures. The Weibull bounds of enhanced curve fitting are denoted as m_1, m_2, and m_3. As the loss modulus was found to be slight as compared to the storage one, it was assumed to be the same as Young's modulus.

Furthermore, Qi et al. (2008) provided the heat-reliant $G_m(T)$ and Poisson's ratio $\mu_m(T)$ as shown in Eqs. 4.154 and 4.155:

$$G_m(T) = \frac{E_m(T)}{2[1+\mu_m(T)]} \tag{4.154}$$

$$\mu_m(T) = \mu_g v_g(T) + \mu_r[1-v_g(T)] \tag{4.155}$$

The heat-reliant $G_m(T)$ and Poisson's ratio ($\mu_m(T)$) were given by Qi et al. (2008). Comparison of the frozen and dynamic Poisson's ratios is represented by μ_g and μ_r, respectively, while v_g is the volume portion of the frozen phase. The values of $v_g(T)$ were obtained using Eq. 4.156:

$$v_g(T) = 1 - \frac{1}{1+\exp\left[-\dfrac{T-T_m}{Z}\right]} \tag{4.156}$$

The current study assumes that the elastic modulus of the SMPC is found by the Rule of Mixtures, which accounts for the fact that the SMP matrix contains fibers. The elastic modulus is represented as E_t and given by Eq. 4.157:

$$E_t(T) = E_f V_f + E_m(T) V_m \tag{4.157}$$

where V_f and V_m are the volume portions of fiber and matrix, respectively, and E_f is the elastic modulus of the fibers. Although previous studies have used Finite element method (FEM) to investigate the thermomechanical properties of SMPC, the present study focuses on presenting an analysis established on a simplified model. The parameters obtained from previous equations were used to estimate the governing equations for shear and longitudinal stress, given by Eqs. 4.158–4.160:

$$\tau_{xy} = G_m(T)\gamma_{xy} \tag{4.158}$$

$$\tau_{yz} = G_m(T)\gamma_{yz} \tag{4.159}$$

$$\sigma_{xx} = E_t(T)\varepsilon_{xx} \tag{4.160}$$

4.4.1 TEMPERATURE-REGULATED PROPERTIES OF SMPC

Figure 4.6a and b depicts the temperature-dependent elastic modulus ($E_m(T)$) and shear modulus ($G_m(T)$) of the SMP matrix, which were calculated using Eqs. 4.25–4.28 and are based on factors listed in Table 4.4. These equations were derived from the analytical investigations conducted by Gu et al. (2019). The flow in Figure 4.14a and b illustrates the dynamic material properties of the SMP matrix in the glass transition region. The shear and elastic moduli remain constant during the initial stage of temperature

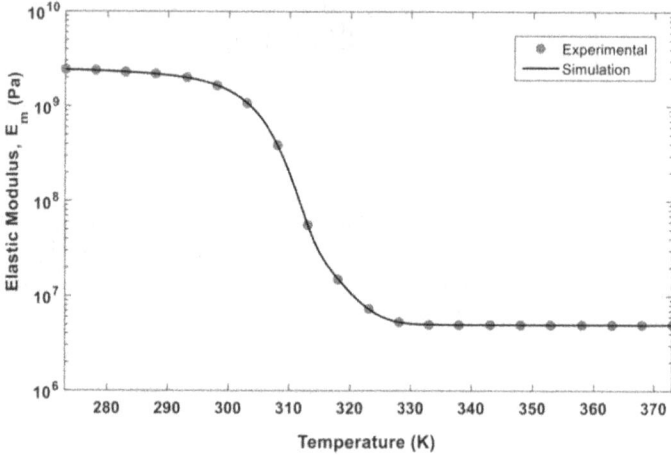

(a)

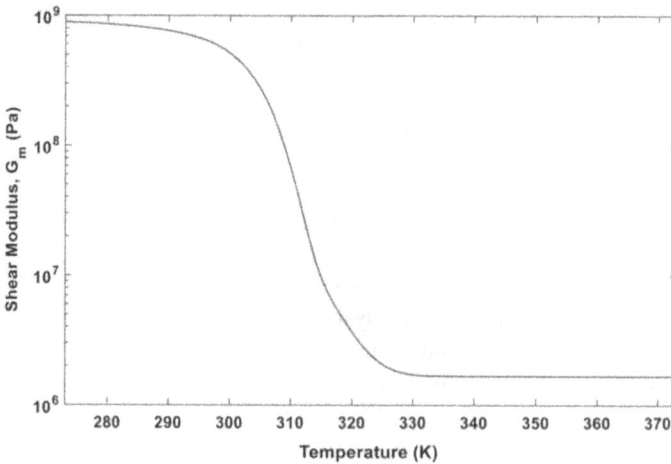

(b)

FIGURE 4.14   Modulus in the glass transition region: (a) Elastic modulus (b) Shear modulus of SMP.

elevation, indicating the elastic behavior of the material. However, during the changes from the starting of the glass transition temperatures to the moving temperature, the modulus declines before stabilizing to a constant value after the transition. This significant conversion in material properties governs the bending behavior of the respective composites, which is further discussed to analyze the effect of temperature increase on microbuckling of fibers caused by bending. Table 4.4 summarizes the values of the parameters considered to model the SMP matrix, as reported by Gu et al. (2019).

SMPC to calculate the elastic modulus in the glass transition temperature range, Eq. 4.29 was used, along with the characteristics of the SMP matrix and the characteristics of reinforced carbon fiber. For carbon fiber the longitudinal modulus was assumed to be 276 GPa, and 0.4 is the set volume fraction to validate the results of the present study. These temperature-dependent properties of the SMPC were then utilized to calculate the strain energy in the plates of the SMPC, the microbuckling characteristics of the fibers, and the positions of the SMPC strain surface in subsequent stages. For microbuckling of the fibers at elevated temperature for SMPC, the parameters used to develop the model are presented in Table 4.5 (Zhang, Dui, and Liang 2018).

Figure 4.15 highlights the impact of temperature by substituting the characteristics of the temperature-controlled SMP matrix. The plots for different temperatures, namely $T = 273$ K, $T_\beta = 295.2$ K, $T_g = 305$ K, and $T_f = 415.5$ K, are shown while considering $k = 50$ m^{-1}. The reduction in the total strain energy during bending of the SMPC with increasing temperature is because of the softening of the SMP matrix at higher temperatures. This softening of the matrix is consistent with the change in moduli of the SMP observed in its glass transition region.

Figure 4.16 depicts the correlation between the total strain energy (U) and temperature change for values of k equal to 5, 10, 15, and 20 m^{-1}.

TABLE 4.4
Values of the Parameters to Model the SMP (Gu et al. 2019)

Parameter	T_g	T_m	T_β	T_f	E_1	E_2	E_3	m_1, m_2, m_3	μ_g	μ_r	Z
Values	305	300.5	295.2	415.5	2.552	1.876	5	19.3,58.4,177.6	0.35	0.49	7
Unit	K	K	K	K	GPa	GPa	GPa	–	–	–	–

SMP, shape memory polymer.

TABLE 4.5
Material and Geometry Parameters for the Microbuckling Calculations (Zhang, Dui, and Liang 2018)

Parameter	Fiber Diameter h	Fiber Volume Fraction	Plate Thickness t	Plate Width w	Plate Length l	Longitudinal Modulus of the Fibers E_f	Half Wavelength λ
Values	7×10^{-3}	0.2	2	5	30	230×10^3	1.25
Unit	mm	–	mm	mm	mm	MPa	mm

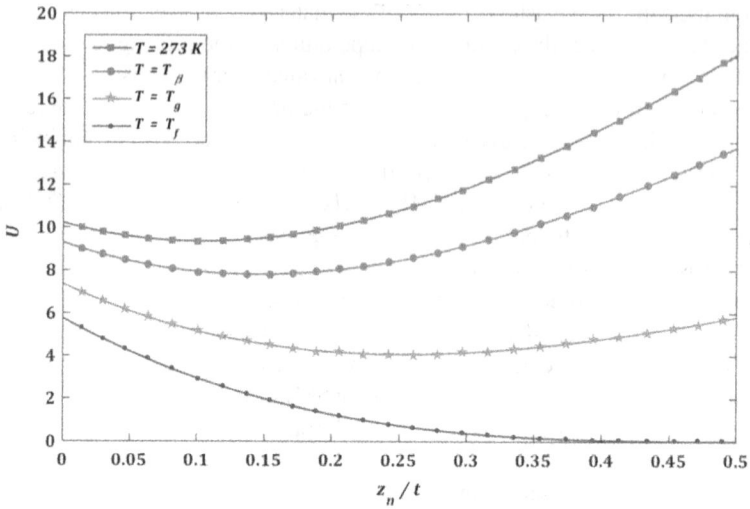

FIGURE 4.15 Strain energy with respect to z_n/t throughout the glass transition region for the SMP matrix (Tiwari and Shaikh 2021c).

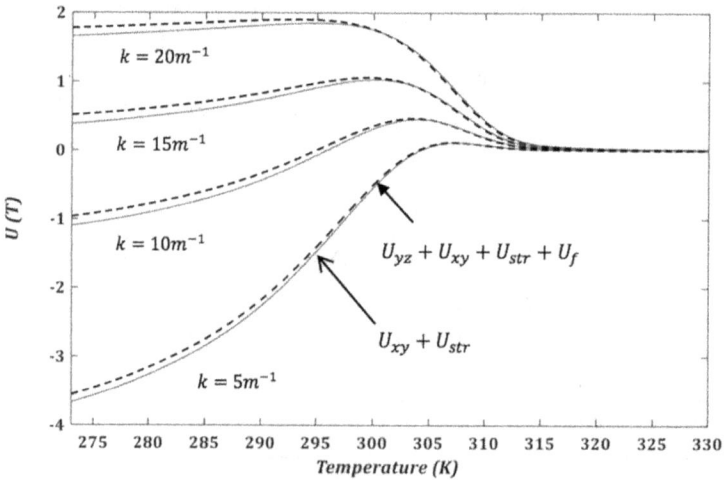

FIGURE 4.16 Change of strain energy with temperature and comparison of total strain energy and sum of U_{xy} and U_{str} (Tiwari and Shaikh 2021c).

4.4.2 Position of Neutral Strain Plane and Critical Buckling Plane

The placement of the neutral strain plane (z_n) and critical buckling plane (z_b) is crucial in determining the positions of buckled fibers during SMPC plate bending. Figures 4.17 and 4.18 illustrate the behavior of z_n/t and z_b/t with respect to temperature (T) for various plots, namely $k = 5$, 10, 20, and 50 m⁻¹. As the bending curves increased, the positions of z_n and z_b shifted in the tensile zone of the composite plate loaded under bending. Once z_n and z_b/t values exceeded zero, the fibers in the SMPC began to micro-buckle. Consequently, as the bending curves increased, the nature of fibers to micro-buckle also enhanced. Furthermore, a rise in temperature in the glass transition region led to a decrease in the moduli of the SMP matrix, which, in turn, increased the tendency of fibers to buckle.

The findings of this study on the positions of z_n and z_b align closely with those reported by Zhang, Dui, and Liang (2018). The curves in Figures 4.17 and 4.18

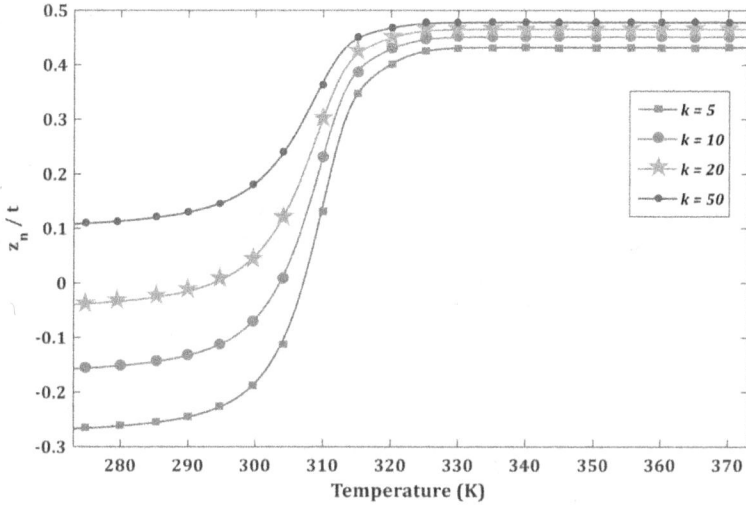

FIGURE 4.17   Change in position of the neutral strain surface z_n/t with respect to temperature under consideration of various degrees of bending, k (m^{-1}) (Tiwari and Shaikh 2021c).

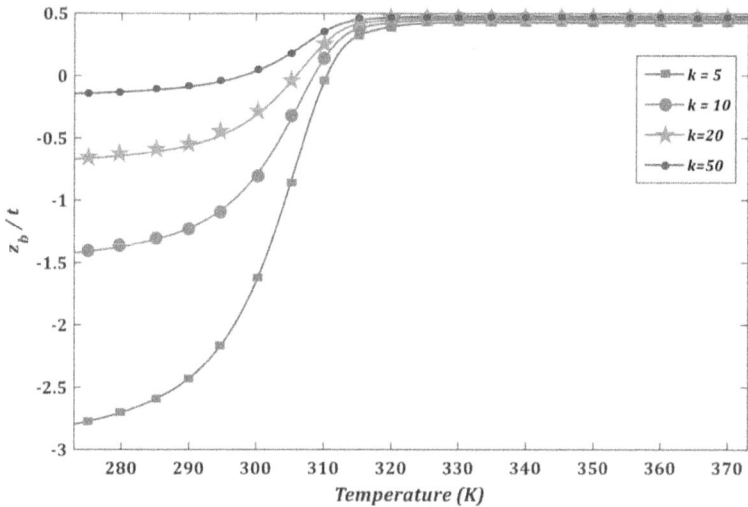

FIGURE 4.18   Position variation of the critical buckling surface z_b/t with temperature for various degrees of bending, k (m^{-1}) (Tiwari and Shaikh 2021c).

trace the total glass transition region of the SMP matrix and provide a comprehensive understanding of microbuckling of fibers in SMPC plates under bending. This interpretation is superior to those of earlier studies that only considered elevated temperatures. This study highlights the crucial role of temperature in altering the location of buckling fibers in SMPCs under high bending deformations.

4.4.3 HALF WAVELENGTH OF THE FIBER

Figure 4.12 illustrates the impact of the fiber volume fraction on λ at three temperatures: $T_\beta = 295.2$ K, $T_g = 305$ K, and $T_f = 415.5$ K. The maximum value of wavelength was observed near a volume fraction of 0.4, which is the same for all temperatures. As the temperature increased, the magnitude of the wavelength also increased due to the decline in the SMP modulus.

In Figure 4.20, the impact of temperature and SMPC plate thickness on the half wavelength is presented for thickness values ranging from 0.5 to 2.5 mm. The results show that the half wavelength increases consistently as the thickness of the plate increases. This increase can be attributed to the larger gap between adjacent fibers at higher thickness values. The analysis was conducted at three temperatures, $T_\beta = 295.2$ K, $T_g = 305$ K, and $T_f = 415.5$ K.

The impact of temperature on the values of λ, for the same deflection in the glass transition region for the SMP matrix is shown in Figure 4.21 for the thickness range of SMPC of 0.5–2 mm. It is considered that the impact of thickness on the wavelength is enhanced at elevated temperatures outside the glass transition region of SMP, as related to it below T_g.

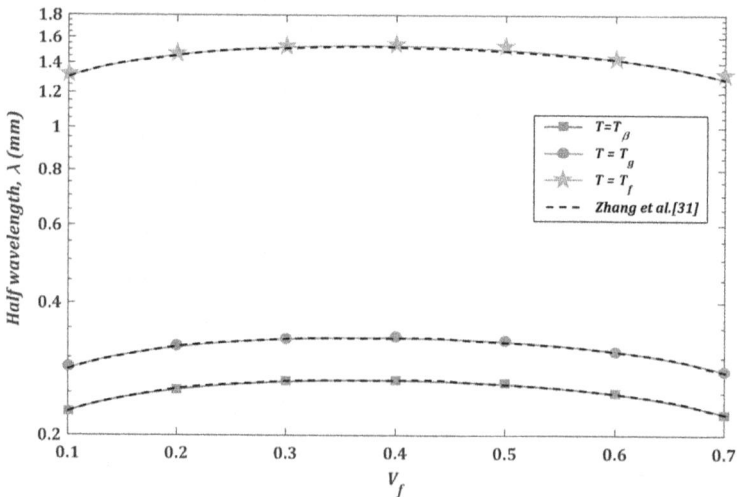

FIGURE 4.19 Change in the half fiber buckling wavelength (λ) with respect to volume fraction of fiber at $T = T_\beta, T_g$ and T_f.

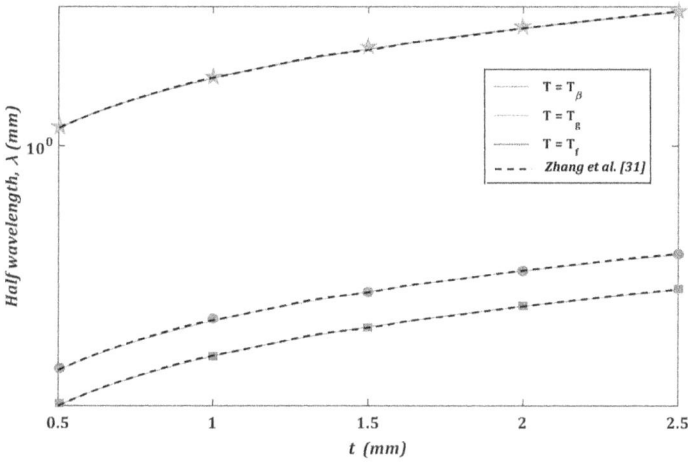

FIGURE 4.20 Change in the half wavelength (λ) with thickness (t) at $T = T_\beta$, T_g and T_f.

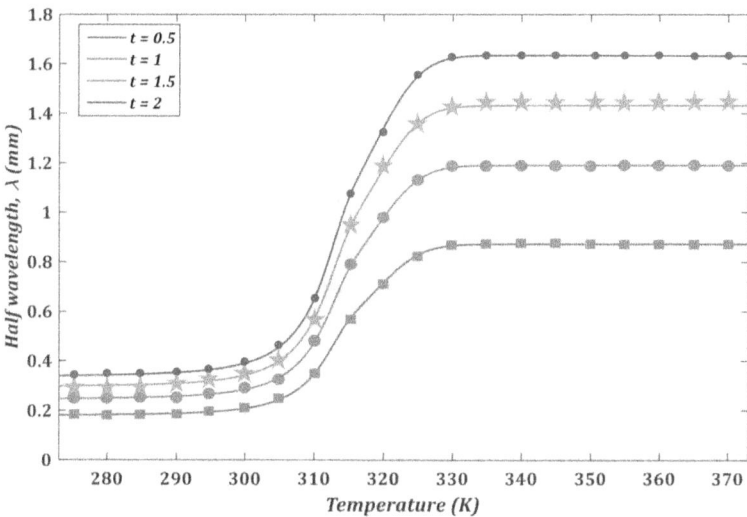

FIGURE 4.21 Change in the half wavelength (λ) with temperature (T) at t from 0.5 to 2 mm.

REFERENCES

Abishera, Rajkumar, Ramraj Velmurugan, and K. V. Nagendra Gopal. 2017. "Reversible Plasticity Shape Memory Effect in Epoxy/CNT Nanocomposites - A Theoretical Study." *Composites Science and Technology* 141 (March): 145–53. https://doi.org/10.1016/j.compscitech.2017.01.020.

Abrahamson, Erik R., Mark S. Lake, Naseem A. Munshi, and Ken Gall. 2003. "Shape Memory Mechanics of an Elastic Memory Composite Resin." *Journal of Intelligent Material Systems and Structures* 14 (10): 623–32. https://doi.org/10.1177/104538903036213.

Arrieta, Sebastián, Julie Diani, and Pierre Gilormini. 2014. "Experimental Characterization and Thermoviscoelastic Modeling of Strain and Stress Recoveries of an Amorphous Polymer Network." *Mechanics of Materials* 68 (January): 95–103. https://doi.org/10.1016/j.mechmat.2013.08.008.

Arruda, Ellen M., and Mary C. Boyce. 1993. "A Three-Dimensional Constitutive Model for the Large Stretch Behavior of Rubber Elastic Materials." *Journal of the Mechanics and Physics of Solids* 41 (2): 389–412. https://doi.org/10.1016/0022-5096(93)90013-6.

Arvanitakis, Antonios I. 2019. "A Constitutive Level-Set Model for Shape Memory Polymers and Shape Memory Polymeric Composites." *Archive of Applied Mechanics* 89 (9): 1939–51. https://doi.org/10.1007/s00419-019-01553-w.

Baghani, Mostafa, Jamal Arghavani, and Reza Naghdabadi. 2014. "A Finite Deformation Constitutive Model for Shape Memory Polymers Based on Hencky Strain." *Mechanics of Materials* 73 (June): 1–10. https://doi.org/10.1016/j.mechmat.2013.11.011.

Baghani, Mostafa, Reza Naghdabadi, Jamal Arghavani, and Saeed Sohrabpour. 2012. "A Thermodynamically-Consistent 3D Constitutive Model for Shape Memory Polymers." *International Journal of Plasticity* 35 (August): 13–30. https://doi.org/10.1016/j.ijplas.2012.01.007.

Baghani, Mostafa, Reza Naghdabadi, Jamal Arghavani, and Saeed Sohrabpour. 2012. "A Constitutive Model for Shape Memory Polymers with Application to Torsion of Prismatic Bars." *Journal of Intelligent Material Systems and Structures* 23 (2): 107–16. https://doi.org/10.1177/1045389X11431745.

Balogun, Olaniyi, and Changki Mo. 2014. "Shape Memory Polymers: Three-Dimensional Isotropic Modeling." *Smart Materials and Structures* 23 (4): 045008. https://doi.org/10.1088/0964-1726/23/4/045008.

Barot, G., and I. Joga Rao. 2006. "Constitutive Modeling of the Mechanics Associated with Crystallizable Shape Memory Polymers." *Zeitschrift Für Angewandte Mathematik Und Physik ZAMP* 57 (4): 652–81. https://doi.org/10.1007/s00033-005-0009-6.

Barot, G., I. Joga Rao, and K. R. Rajagopal. 2008. "A Thermodynamic Framework for the Modeling of Crystallizable Shape Memory Polymers." *International Journal of Engineering Science* 46 (4): 325–51. https://doi.org/10.1016/j.ijengsci.2007.11.008.

Bhattacharyya, A., and H. Tobushi. 2000. "Analysis of the Isothermal Mechanical Response of a Shape Memory Polymer Rheological Model." *Polymer Engineering & Science* 40 (12): 2498–510. https://doi.org/10.1002/pen.11381.

Boatti, Elisa, Giulia Scalet, and Ferdinando Auricchio. 2016. "A Three-Dimensional Finite-Strain Phenomenological Model for Shape-Memory Polymers: Formulation, Numerical Simulations, and Comparison with Experimental Data." *International Journal of Plasticity* 83 (August): 153–77. https://doi.org/10.1016/j.ijplas.2016.04.008.

Bonner, M., H. Montes de Oca, M. Brown, and I. M. Ward. 2010. "A Novel Approach to Predict the Recovery Time of Shape Memory Polymers." *Polymer* 51 (6): 1432–6. https://doi.org/10.1016/j.polymer.2010.01.058.

Bouaziz, R., F. Roger, and K. Prashantha. 2017. "Thermo-Mechanical Modeling of Semi-Crystalline Thermoplastic Shape Memory Polymer under Large Strain." *Smart Materials and Structures* 26 (5): 055009. https://doi.org/10.1088/1361-665X/aa6690.

Castro, Francisco, Kristofer K. Westbrook, Kevin N. Long, Robin Shandas, and H. Jerry Qi. 2010. "Effects of Thermal Rates on the Thermomechanical Behaviors of Amorphous Shape Memory Polymers." *Mechanics of Time-Dependent Materials* 14 (3): 219–41. https://doi.org/10.1007/s11043-010-9109-6.

Chamis, Christos C. 1983. "*Simplified Composite Micromechanics Equations for Hygral, Thermal and Mechanical Properties.*" Houston, TX. https://ntrs.nasa.gov/citations/19830011546.

Chamis, Christos C., and G. P. Sendeckyj. 1968. "Critique on Theories Predicting Thermoelastic Properties of Fibrous Composites." *Journal of Composite Materials* 2 (3): 332–58. https://doi.org/10.1177/002199836800200305.

Chamis, Christos. 1969. *Failure Criteria for Filamentary Composites*. National Aeronautics and Space Administration, Washington, DC.

Chen, Xiang, and Thao. D. Nguyen. 2011. "Influence of Thermoviscoelastic Properties and Loading Conditions on the Recovery Performance of Shape Memory Polymers." *Mechanics of Materials* 43 (3): 127–38. https://doi.org/10.1016/j.mechmat.2011.01.001.

Chen, Yi-Chao, and Dimitris C. Lagoudas. 2008a. "A Constitutive Theory for Shape Memory Polymers. Part I: Large Deformations." *Journal of the Mechanics and Physics of Solids* 56 (5): 1752–65. https://doi.org/10.1016/j.jmps.2007.12.005.

Chen, Yi-Chao, and Dimitris C. Lagoudas. 2008b. "A Constitutive Theory for Shape Memory Polymers. Part II: A Linearized Model for Small Deformations." *Journal of the Mechanics and Physics of Solids* 56 (5): 1766–8. https://doi.org/10.1016/j.jmps.2007.12.004.

Choi, Jinwoo, Alicia M. Ortega, Rui Xiao, Christopher M. Yakacki, and Thao D. Nguyen. 2012. "Effect of Physical Aging on the Shape-Memory Behavior of Amorphous Networks." *Polymer* 53 (12): 2453–64. https://doi.org/10.1016/j.polymer.2012.03.066.

Diani, Julie, Yiping Liu, and Ken Gall. 2006. "Finite Strain 3D Thermoviscoelastic Constitutive Model for Shape Memory Polymers." *Polymer Engineering & Science* 46 (4): 486–92. https://doi.org/10.1002/pen.20497.

Dong, Yubing, Yaofeng Zhu, Meng Liu, Qinxi Dong, Ran Li, and Yaqin Fu. 2018. "Constitutive Model for Shape Memory Polyurethane Based on Phase Transition and One-Dimensional Non-Linear Viscoelastic." *Materials Today Communications* 17 (December): 133–9. https://doi.org/10.1016/j.mtcomm.2018.08.020.

Eshelby, John Douglas. 1957. "The Determination of the Elastic Field of an Ellipsoidal Inclusion, and Related Problems." Proceedings of the Royal Society of London. *Series A. Mathematical and Physical Sciences* 241 (1226): 376–96. https://doi.org/10.1098/rspa.1957.0133.

Fan, Pengxuan, Wujun Chen, Bing Zhao, Jianhui Hu, Jifeng Gao, Guangqiang Fang, and Fujun Peng. 2018. "Formulation and Numerical Implementation of Tensile Shape Memory Process of Shape Memory Polymers." *Polymer* 148 (July): 370–81. https://doi.org/10.1016/j.polymer.2018.06.054.

Fang, Changqing, Jinsong Leng, Huiyu Sun, and Jianping Gu. 2018. "A Multi-Branch Thermoviscoelastic Model Based on Fractional Derivatives for Free Recovery Behaviors of Shape Memory Polymers." *Mechanics of Materials* 120 (May): 34–42. https://doi.org/10.1016/j.mechmat.2018.03.002.

Fang, C.-Q., H.-Y. Sun, and J.-P. Gu. 2016. "A Fractional Calculus Approach to the Prediction of Free Recovery Behaviors of Amorphous Shape Memory Polymers." *Journal of Mechanics* 32 (1): 11–7. https://doi.org/10.1017/jmech.2015.82.

Francis, William, Mark Lake, Marc Schultz, Douglas Campbell, Martin Dunn, and H. Jerry Qi. 2012. "Elastic Memory Composite Microbuckling Mechanics: Closed-Form Model with Empirical Correlation." In *48th AIAA/ASME/ASCE/AHS/ASC Structures, Structural Dynamics, and Materials Conference*. American Institute of Aeronautics and Astronautics. https://doi.org/10.2514/6.2007-2164.

Ge, Qi, Xiaofan Luo, Erika D. Rodriguez, Xiao Zhang, Patrick T. Mather, Martin L. Dunn, and H. Jerry Qi. 2012. "Thermomechanical Behavior of Shape Memory Elastomeric Composites." *Journal of the Mechanics and Physics of Solids* 60 (1): 67–83. https://doi.org/10.1016/j.jmps.2011.09.011.

Ge, Qi, Ahmad Serjouei, H. Jerry Qi, and Martin L. Dunn. 2016. "Thermomechanics of Printed Anisotropic Shape Memory Elastomeric Composites." *International Journal of Solids and Structures* 102-103 (December): 186–99. https://doi.org/10.1016/j.ijsolstr.2016.10.005.

Ge, Qi, Kai Yu, Yifu Ding, and H. Jerry Qi. 2012. "Prediction of Temperature-Dependent Free Recovery Behaviors of Amorphous Shape Memory Polymers." *Soft Matter* 8 (43): 11098–105. https://doi.org/10.1039/C2SM26249E.

Ghosh, Pritha, and A. R. Srinivasa. 2011. "A Two-Network Thermomechanical Model of a Shape Memory Polymer." *International Journal of Engineering Science* 49 (9): 823–38. https://doi.org/10.1016/j.ijengsci.2011.04.003.

Ghosh, Pritha, and A. R. Srinivasa. 2013. "A Two-Network Thermomechanical Model and Parametric Study of the Response of Shape Memory Polymers." *Mechanics of Materials* 60 (July): 1–17. https://doi.org/10.1016/j.mechmat.2012.12.005.

Gilormini, Pierre, and Julie Diani. 2012. "On Modeling Shape Memory Polymers as Thermoelastic Two-Phase Composite Materials." *Comptes Rendus Mécanique*, Recent Advances in Micromechanics of Materials, 340 (4): 338–48. https://doi.org/10.1016/j.crme.2012.02.016.

Gu, Jianping, Jinsong Leng, and Huiyu Sun. 2017. "A Constitutive Model for Amorphous Shape Memory Polymers Based on Thermodynamics with Internal State Variables." *Mechanics of Materials* 111 (August): 1–14. https://doi.org/10.1016/j.mechmat.2017.04.008.

Gu, Jianping, Jinsong Leng, Huiyu Sun, Hao Zeng, and Zhongbing Cai. 2019. "Thermomechanical Constitutive Modeling of Fiber Reinforced Shape Memory Polymer Composites Based on Thermodynamics with Internal State Variables." *Mechanics of Materials* 130 (March): 9–19. https://doi.org/10.1016/j.mechmat.2019.01.004.

Gu, Jianping, Huiyu Sun, and Changqing Fang. 2014. "A Multi-Branch Finite Deformation Constitutive Model for a Shape Memory Polymer Based Syntactic Foam." *Smart Materials and Structures* 24 (2): 025011. https://doi.org/10.1088/0964-1726/24/2/025011.

Gu, Jianping, Huiyu Sun, and Changqing Fang. 2015. "A Phenomenological Constitutive Model for Shape Memory Polyurethanes." *Journal of Intelligent Material Systems and Structures* 26 (5): 517–26. https://doi.org/10.1177/1045389X14530595.

Gu, Jianping, Huiyu Sun, Jianshi Fang, Changqing Fang, and Zhenqin Xu. 2016. "A Unified Modeling Approach for Amorphous Shape Memory Polymers and Shape Memory Polymer Based Syntactic Foam." *Polymers for Advanced Technologies* 27 (9): 1237–45. https://doi.org/10.1002/pat.3789.

Guo, Jianming, Jingbiao Liu, Zhenqing Wang, Xiaofu He, Lifeng Hu, Liyong Tong, and Xiaojun Tang. 2017. "A Thermodynamics Viscoelastic Constitutive Model for Shape Memory Polymers." *Journal of Alloys and Compounds* 705 (May): 146–55. https://doi.org/10.1016/j.jallcom.2017.02.142.

Guo, Xiaogang, Liwu Liu, Yanju Liu, Bo Zhou, and Jinsong Leng. 2014. "Constitutive Model for a Stress- and Thermal-Induced Phase Transition in a Shape Memory Polymer." *Smart Materials and Structures* 23 (10): 105019. https://doi.org/10.1088/0964-1726/23/10/105019.

Guo, Xiaogang, Liwu Liu, Bo Zhou, Yanju Liu, and Jinsong Leng. 2015. "Influence of Strain Rates on the Mechanical Behaviors of Shape Memory Polymer." *Smart Materials and Structures* 24 (9): 095009. https://doi.org/10.1088/0964-1726/24/9/095009.

Guo, Xiaogang, Liwu Liu, Bo Zhou, Yanju Liu, and Jinsong Leng. 2016. "Constitutive Model for Shape Memory Polymer Based on the Viscoelasticity and Phase Transition Theories." *Journal of Intelligent Material Systems and Structures* 27 (3): 314–23. https://doi.org/10.1177/1045389X15571380.

Hahn, H. T. 1980. "Simplified Formulas for Elastic Moduli of Unidirectional Continuous Fiber Composites." *Composites Technology and Research* 2 (3): 5–7. https://doi.org/10.1520/CTR10702J.

Halpin, John C. 1969. *"Effects of Environmental Factors on Composite Materials."* Air Force Materials Lab Wright-Patterson AFB OH.

Hashin, Zvi, and B. Walter Rosen. 1964. "The Elastic Moduli of Fiber-Reinforced Materials." *Journal of Applied Mechanics* 31 (2): 223–32. https://doi.org/10.1115/1.3629590.

Haupt, Peter, Alexander Lion, and E. Backhaus. 2000. "On the Dynamic Behaviour of Polymers under Finite Strains: Constitutive Modelling and Identification of Parameters." *International Journal of Solids and Structures* 37 (26): 3633–46. https://doi.org/10.1016/S0020-7683(99)00165-1.

Heuchel, Matthias., Junfeng Cui, K. Kratz, Hans Kosmella, and Andreas Lendlein. 2010. "Relaxation Based Modeling of Tunable Shape Recovery Kinetics Observed under Isothermal Conditions for Amorphous Shape-Memory Polymers." *Polymer* 51 (26): 6212–8. https://doi.org/10.1016/j.polymer.2010.10.051.

Hill, R. 1964a. "Theory of Mechanical Properties of Fibre-Strengthened Materials: I. Elastic Behaviour." *Journal of the Mechanics and Physics of Solids* 12 (4): 199–212. https://doi.org/10.1016/0022-5096(64)90019-5.

Hill, R. 1964b. "Theory of Mechanical Properties of Fibre-Strengthened Materials: II. Inelastic Behaviour." *Journal of the Mechanics and Physics of Solids* 12 (4): 213–8. https://doi.org/10.1016/0022-5096(64)90020-1.

Holzapfel, Gerhard A. 2002. "Nonlinear Solid Mechanics: A Continuum Approach for Engineering Science." *Meccanica* 37 (4): 489–90. https://doi.org/10.1023/A:1020843529530.

Khajehsaeid, H., J. Arghavani, R. Naghdabadi, and S. Sohrabpour. 2014. "A Visco-Hyperelastic Constitutive Model for Rubber-like Materials: A Rate-Dependent Relaxation Time Scheme." *International Journal of Engineering Science* 79 (June): 44–58. https://doi.org/10.1016/j.ijengsci.2014.03.001.

Khonakdar, Hossein Ali, Seyed Hassan Jafari, Sorour Rasouli, Jalil Morshedian, and Hossein Abedini. 2007. "Investigation and Modeling of Temperature Dependence Recovery Behavior of Shape-Memory Crosslinked Polyethylene." *Macromolecular Theory and Simulations* 16 (1): 43–52. https://doi.org/10.1002/mats.200600041.

Kim, Ju Hyun, Tae Jin Kang, and Woong-Ryeol Yu. 2010. "Thermo-Mechanical Constitutive Modeling of Shape Memory Polyurethanes Using a Phenomenological Approach." *International Journal of Plasticity* 26 (2): 204–18. https://doi.org/10.1016/j.ijplas.2009.06.006.

Kolesov, Igor, Oleksandr Dolynchuk, and Hans Joachim Radusch. 2013. "Modeling of Shape-Memory Recovery in Crosslinked Semicrystalline Polymers." *Advances in Science and Technology* 77: 319–24. https://doi.org/10.4028/www.scientific.net/AST.77.319.

Lan, Xin, Liwu Liu, Yanju Liu, and Jinsong Leng. 2018. "Thermomechanical and Electroactive Behavior of a Thermosetting Styrene-Based Carbon Black Shape-Memory Composite." *Journal of Applied Polymer Science* 135 (13): 45978. https://doi.org/10.1002/app.45978.

Leng, Jinsong, Xin Lan, Yanju Liu, and Shanyi Du. 2011. "Shape-Memory Polymers and Their Composites: Stimulus Methods and Applications." *Progress in Materials Science* 56 (7): 1077–135. https://doi.org/10.1016/j.pmatsci.2011.03.001.

Li, Guoqiang, and Damon Nettles. 2010. "Thermomechanical Characterization of a Shape Memory Polymer Based Self-Repairing Syntactic Foam." *Polymer* 51 (3): 755–62. https://doi.org/10.1016/j.polymer.2009.12.002.

Li, Guoqiang, and Wei Xu. 2011. "Thermomechanical Behavior of Thermoset Shape Memory Polymer Programmed by Cold-Compression: Testing and Constitutive Modeling." *Journal of the Mechanics and Physics of Solids* 59 (6): 1231–50. https://doi.org/10.1016/j.jmps.2011.03.001.

Li, Wenbing, Yanju Liu, and Jinsong Leng. 2015. "Selectively Actuated Multi-Shape Memory Effect of a Polymer Multicomposite." *Journal of Materials Chemistry A* 3 (48): 24532–9. https://doi.org/10.1039/C5TA08513F.

Li, Yunxin, Siu-Siu Guo, Yuhao He, and Zishun Liu. 2015. "A Simplified Constitutive Model for Predicting Shape Memory Polymers Deformation Behavior." *International Journal of Computational Materials Science and Engineering* 04 (01): 1550001. https://doi.org/10.1142/S2047684115500013.

Li, Yunxin, Yuhao He, and Zishun Liu. 2017. "A Viscoelastic Constitutive Model for Shape Memory Polymers Based on Multiplicative Decompositions of the Deformation Gradient." *International Journal of Plasticity* 91 (April): 300–17. https://doi.org/10.1016/j.ijplas.2017.04.004.

Li, Yunxin, Jianying Hu, and Zishun Liu. 2017. "A Constitutive Model of Shape Memory Polymers Based on Glass Transition and the Concept of Frozen Strain Release Rate." *International Journal of Solids and Structures* 124 (October): 252–63. https://doi.org/10.1016/j.ijsolstr.2017.06.039.

Li, Yunxin, and Zishun Liu. 2018. "A Novel Constitutive Model of Shape Memory Polymers Combining Phase Transition and Viscoelasticity." Polymer 143 (May): 298–308. https://doi.org/10.1016/j.polymer.2018.04.026.

Li, Zheng, Hong Wang, Rui Xiao, and Su Yang. 2017. "A Variable-Order Fractional Differential Equation Model of Shape Memory Polymers." *Chaos, Solitons & Fractals, Future Directions in Fractional Calculus Research and Applications* 102 (September): 473–85. https://doi.org/10.1016/j.chaos.2017.04.042.

Lin, J. R., and Liang-Wei Chen. 1999. "Shape-Memorized Crosslinked Ester-Type Polyurethane and Its Mechanical Viscoelastic Model." *Journal of Applied Polymer Science* 73 (7): 1305–19. https://doi.org/10.1002/(SICI)1097-4628(19990815)73:7<1305::AID-APP24>3.0.CO;2-5.

Liu, Yiping, Ken Gall, Martin L. Dunn, Alan R. Greenberg, and Julie Diani. 2006. "Thermomechanics of Shape Memory Polymers: Uniaxial Experiments and Constitutive Modeling." *International Journal of Plasticity* 22 (2): 279–313. https://doi.org/10.1016/j.ijplas.2005.03.004.

Lu, Haibao, Xiaodong Wang, Yongtao Yao, and Yong Qing Fu. 2018. "A 'Frozen Volume' Transition Model and Working Mechanism for the Shape Memory Effect in Amorphous Polymers." *Smart Materials and Structures* 27 (6): 065023. https://doi.org/10.1088/1361-665X/aab8af.

Luo, Ling, Yu Xi Jia, Xiao Xia Wang, Zhao Jing Wang, Jun Peng Gao, and Xiao Su Yi. 2014. "Modeling of Polyurethane's Shape Memory Effects and Recovery Time under Different Recovery Temperatures." *Key Engineering Materials* 575-576: 95–100. https://doi.org/10.4028/www.scientific.net/KEM.575-576.95.

McClung, Amber J. W., Gyaneshwar P. Tandon, and Jeffery W. Baur. 2013. "Deformation Rate-, Hold Time-, and Cycle-Dependent Shape-Memory Performance of Veriflex-E Resin." *Mechanics of Time-Dependent Materials* 17 (1): 39–52. https://doi.org/10.1007/s11043-011-9157-6.

Mori, Trevor A., and Ken Tanaka. 1973. "Average Stress in Matrix and Average Elastic Energy of Materials with Misfitting Inclusions." *Acta Metallurgica* 21 (5): 571–4. https://doi.org/10.1016/0001-6160(73)90064-3.

Morshedian, Jalil, Hossein A. Khonakdar, and Sorour Rasouli. 2005. "Modeling of Shape Memory Induction and Recovery in Heat-Shrinkable Polymers." *Macromolecular Theory and Simulations* 14 (7): 428–34. https://doi.org/10.1002/mats.200400108.

Nguyen, Thao D. 2013. "Modeling Shape-Memory Behavior of Polymers." *Polymer Reviews* 53 (1): 130–52. https://doi.org/10.1080/15583724.2012.751922.

Nguyen, Thao D., H. Jerry Qi, Francisco Castro, and Kevin N. Long. 2008. "A Thermoviscoelastic Model for Amorphous Shape Memory Polymers: Incorporating Structural and Stress Relaxation." *Journal of the Mechanics and Physics of Solids* 56 (9): 2792–814. https://doi.org/10.1016/j.jmps.2008.04.007.

Nguyen, Thao D., Christopher M. Yakacki, Parth D. Brahmbhatt, and Matthew L. Chambers. 2010. "Modeling the Relaxation Mechanisms of Amorphous Shape Memory Polymers." *Advanced Materials* 22 (31): 3411–23. https://doi.org/10.1002/adma.200904119.

Odegard, Gregory, Thomas Gates, Kristopher Wise, Cheol Park, and Emilie Siochi. 2003. "Constitutive Modeling of Nanotube-Reinforced Polymer Composites." *Composites Science and Technology, Modeling and Characterization of Nanostructured Materials*, 63 (11): 1671–87. https://doi.org/10.1016/S0266-3538(03)00063-0.

Pan, Zhouzhou, Rong Huang, and Zishun Liu. 2019. "Prediction of the Thermomechanical Behavior of Particle Reinforced Shape Memory Polymers." *Polymer Composites* 40 (1): 353–63. https://doi.org/10.1002/pc.24658.

Pan, Zhouzhou, and Zishun Liu. 2018. "A Novel Fractional Viscoelastic Constitutive Model for Shape Memory Polymers." *Journal of Polymer Science Part B: Polymer Physics* 56 (16): 1125–34. https://doi.org/10.1002/polb.24631.

Pan, Zhouzhou, Yu Zhou, Ni Zhang, and Zishun Liu. 2018. "A Modified Phase-Based Constitutive Model for Shape Memory Polymers." *Polymer International* 67 (12): 1677–83. https://doi.org/10.1002/pi.5698.

Park, Haedong, Philip Harrison, Zaoyang Guo, Myoung-Gue Lee, and Woong-Ryeol Yu. 2016. "Three-Dimensional Constitutive Model for Shape Memory Polymers Using Multiplicative Decomposition of the Deformation Gradient and Shape Memory Strains." *Mechanics of Materials* 93 (February): 43–62. https://doi.org/10.1016/j. mechmat.2015.10.014.

Pieczyska, E. A., M. Maj, K. Kowalczyk-Gajewska, M. Staszczak, A. Gradys, M. Majewski, M. Cristea, H. Tobushi, and S. Hayashi. 2015. "Thermomechanical Properties of Polyurethane Shape Memory Polymer-Experiment and Modelling." *Smart Materials and Structures* 24 (4): 045043. https://doi.org/10.1088/0964-1726/24/4/045043.

Pulla, Sesha S., Mohammad Souri, Haluk E. Karaca, and Y. Charles Lu. 2015. "Characterization and Strain-Energy-Function-Based Modeling of the Thermomechanical Response of Shape-Memory Polymers." *Journal of Applied Polymer Science* 132 (18): 41861–69. https://doi.org/10.1002/app.41861.

Qi, H. Jerry, Thao D. Nguyen, Francisco Castro, Christopher M. Yakacki, and Robin Shandas. 2008. "Finite Deformation Thermo-Mechanical Behavior of Thermally Induced Shape Memory Polymers." *Journal of the Mechanics and Physics of Solids* 56 (5): 1730–51. https://doi.org/10.1016/j.jmps.2007.12.002.

Rajagopal, Kumbakonam Ramamani, and Alan Wineman. 1992. "A Constitutive Equation for Nonlinear Solids Which Undergo Deformation Induced Microstructural Changes." *International Journal of Plasticity* 8 (4): 385–95. https://doi.org/10.1016/0749-6419(92)90056-I.

Reese, Stefanie, Markus Böl, and Daniel Christ. 2010. "Finite Element-Based Multi-Phase Modelling of Shape Memory Polymer Stents." *Computer Methods in Applied Mechanics and Engineering, Multiscale Models and Mathematical Aspects in Solid and Fluid Mechanics*, 199 (21): 1276–86. https://doi.org/10.1016/j.cma.2009.08.014.

Reese, Stefanie, and Sanjay Govindjee. 1998. "A Theory of Finite Viscoelasticity and Numerical Aspects." *International Journal of Solids and Structures* 35 (26): 3455–82. https://doi.org/10.1016/S0020-7683(97)00217-5.

Saleeb, Atef, Sufian Natsheh, and Josiah Owusu-Danquah. 2017. "A Multi-Mechanism Model for Large-Strain Thermomechanical Behavior of Polyurethane Shape Memory Polymer." *Polymer* 130 (November): 230–41. https://doi.org/10.1016/j.polymer.2017.10.003.

Saleeb, Atef, Santo Padula, and A. Kumar. 2011. "A Multi-Axial, Multimechanism Based Constitutive Model for the Comprehensive Representation of the Evolutionary Response of SMAs under General Thermomechanical Loading Conditions." *International Journal of Plasticity* 27 (5): 655–87. https://doi.org/10.1016/j.ijplas.2010.08.012.

Shi, Dong-Li, Xi-Qiao Feng, Yonggang Y. Huang, Keh-Chih Hwang, and Huajian Gao. 2004. "The Effect of Nanotube Waviness and Agglomeration on the Elastic Property of Carbon Nanotube-Reinforced Composites." *Journal of Engineering Materials and Technology* 126 (3): 250–7. https://doi.org/10.1115/1.1751182.

Shi, Guanghui, Qingsheng Yang, Xiaoqiao He, and Kim Meow Liew. 2013. "A Three-Dimensional Constitutive Equation and Finite Element Method Implementation for Shape Memory Polymers." *CMES - Computer Modeling in Engineering and Sciences* 90 (5): 339–58.

Srivastava, Vikas, Shawn A. Chester, and Lallit Anand. 2010. "Thermally Actuated Shape-Memory Polymers: Experiments, Theory, and Numerical Simulations." *Journal of the Mechanics and Physics of Solids* 58 (8): 1100–24. https://doi.org/10.1016/j. jmps.2010.04.004.

Su, Xiaobin, and Xiongqi Peng. 2018. "A 3D Finite Strain Viscoelastic Constitutive Model for Thermally Induced Shape Memory Polymers Based on Energy Decomposition." *International Journal of Plasticity* 110 (November): 166–82. https://doi.org/10.1016/j.ijplas.2018.07.002.

Taherzadeh, M., Mostafa Baghani, Mahmoud Baniassadi, K. Abrinia, and M. Safdari. 2016. "Modeling and Homogenization of Shape Memory Polymer Nanocomposites." *Composites Part B: Engineering* 91 (April): 36–43. https://doi.org/10.1016/j.compositesb.2015.12.044.

Tiwari, Nilesh, and AbdulHafiz A. Shaikh. 2021a. "Buckling and Vibration Analysis of Shape Memory Laminated Composite Beams under Axially Heterogeneous In-Plane Loads in the Glass Transition Temperature Region." *SN Applied Sciences* 3 (4): 1–15.

Tiwari, Nilesh, and AbdulHafiz A. Shaikh. 2021b. "Micro Buckling of Carbon Fiber in Triple Shape Memory Polymer Composites under Bending in Glass Transition Regions." *Materials Today: Proceedings* 44: 4744–8.

Tiwari, Nilesh, and AbdulHafiz A. Shaikh. 2021c. "Micro-Buckling of Carbon Fibers in Shape Memory Polymer Composites under Bending in the Glass Transition Temperature Region." *Curved and Layered Structures* 8 (1): 96–108.

Tobushi, Hisaaki, Takahiro Hashimoto, Shunichi Hayashi, and Etsuko Yamada. 1997. "Thermomechanical Constitutive Modeling in Shape Memory Polymer of Polyurethane Series." *Journal of Intelligent Material Systems and Structures* 8 (8): 711–8. https://doi.org/10.1177/1045389X9700800808.

Tobushi, Hisaaki, Kayo Okumura, Shunichi Hayashi, and Norimitsu Ito. 2001. "Thermomechanical Constitutive Model of Shape Memory Polymer." *Mechanics of Materials* 33 (10): 545–54. https://doi.org/10.1016/S0167-6636(01)00075-8.

Tornabene, Francesco, Nicholas Fantuzzi, Michele Bacciocchi, and Erasmo Viola. 2016. "Effect of Agglomeration on the Natural Frequencies of Functionally Graded Carbon Nanotube-Reinforced Laminated Composite Doubly-Curved Shells." *Composites Part B: Engineering* 89 (March): 187–218. https://doi.org/10.1016/j.compositesb.2015.11.016.

Tsai, Stephen W. 1964. "*Structural Behavior of Composite Materials.*" Philco Corp Newport Beach Ca Space and Re-Entry Systems.

Tsai, Stephen W. 1965. "*Strength Characteristics of Composite Materials.*" Philco Corp Newport Beach CA.

Volk, Brent L., Dimitris C. Lagoudas, and Yi-Chao Chen. 2010. "Analysis of the Finite Deformation Response of Shape Memory Polymers: II. 1D Calibration and Numerical Implementation of a Finite Deformation, Thermoelastic Model." *Smart Materials and Structures* 19 (7): 075006. https://doi.org/10.1088/0964-1726/19/7/075006.

Volk, Brent L., Dimitris C. Lagoudas, Yi-Chao Chen, and Karen S. Whitley. 2010a. "Analysis of the Finite Deformation Response of Shape Memory Polymers: I. Thermomechanical Characterization." *Smart Materials and Structures* 19 (7): 075005. https://doi.org/10.1088/0964-1726/19/7/075005.

Volk, Brent L., Dimitris C. Lagoudas, Yi-Chao Chen, and Karen S. Whitley. 2010b. "Analysis of the Finite Deformation Response of Shape Memory Polymers: I. Thermomechanical Characterization." *Smart Materials and Structures* 19 (7): 075005. https://doi.org/10.1088/0964-1726/19/7/075005.

Volk, Brent L., Dimitris C. Lagoudas, and Duncan J. Maitland. 2011. "Characterizing and Modeling the Free Recovery and Constrained Recovery Behavior of a Polyurethane Shape Memory Polymer." *Smart Materials and Structures* 20 (9): 094004. https://doi.org/10.1088/0964-1726/20/9/094004.

Wang, Yanmei, Yanen Wang, Qinghua Wei, and Juan Zhang. 2022. "Light-Responsive Shape Memory Polymer Composites." *European Polymer Journal* 173 (June): 111314. https://doi.org/10.1016/j.eurpolymj.2022.111314.

Wang, Z. D., D. F. Li, Z. Y. Xiong, and R. N. Chang. 2009. "Modeling Thermomechanical Behaviors of Shape Memory Polymer." *Journal of Applied Polymer Science* 113 (1): 651–6. https://doi.org/10.1002/app.29656.

Westbrook, Kristofer K., Philip H. Kao, Francisco Castro, Yifu Ding, and H. Jerry Qi. 2011. "A 3D Finite Deformation Constitutive Model for Amorphous Shape Memory Polymers: A Multi-Branch Modeling Approach for Nonequilibrium Relaxation Processes." *Mechanics of Materials* 43 (12): 853–69. https://doi.org/10.1016/j.mechmat.2011.09.004.

Westbrook, Kristofer K., Patrick T. Mather, Vikas Parakh, Martin L. Dunn, Qi Ge, Brendan M. Lee, and H. Jerry Qi. 2011. "Two-Way Reversible Shape Memory Effects in a Free-Standing Polymer Composite." *Smart Materials and Structures* 20 (6): 065010. https://doi.org/10.1088/0964-1726/20/6/065010.

Westbrook, Kristofer K., Vikas Parakh, Taekwoong Chung, Patrick T. Mather, Logan C. Wan, Martin L. Dunn, and H. Jerry Qi. 2010. "Constitutive Modeling of Shape Memory Effects in Semicrystalline Polymers With Stretch Induced Crystallization." *Journal of Engineering Materials and Technology* 132 (4): 041010. https://doi.org/10.1115/1.4001964.

Wong, Yokesan San, Z. H. Stachurski, and Subbu S. Venkatraman. 2011. "Modeling Shape Memory Effect in Uncrosslinked Amorphous Biodegradable Polymer." *Polymer* 52 (3): 874–80. https://doi.org/10.1016/j.polymer.2010.12.004.

Xiao, Rui, Christopher M. Yakacki, Jingkai Guo, Carl P. Frick, and Thao D. Nguyen. 2016. "A Predictive Parameter for the Shape Memory Behavior of Thermoplastic Polymers." *Journal of Polymer Science Part B: Polymer Physics* 54 (14): 1405–14. https://doi.org/10.1002/polb.23981.

Xu, We, and Guoqiang Li. 2010. "Constitutive Modeling of Shape Memory Polymer Based Self-Healing Syntactic Foam." *International Journal of Solids and Structures* 47 (9): 1306–16. https://doi.org/10.1016/j.ijsolstr.2010.01.015.

Yang, Qianxi, and Guoqiang Li. 2016. "Temperature and Rate Dependent Thermomechanical Modeling of Shape Memory Polymers with Physics Based Phase Evolution Law." *International Journal of Plasticity* 80 (May): 168–86. https://doi.org/10.1016/j.ijplas.2015.09.005.

Yu, Kai, Hao Li, Amber J. W. McClung, Gyaneshwar P. Tandon, Jeffery W. Baur, and H. Jerry Qi. 2016. "Cyclic Behaviors of Amorphous Shape Memory Polymers." *Soft Matter* 12 (13): 3234–45. https://doi.org/10.1039/C5SM02781K.

Yu, Kai, Kristofer K Westbrook, Philips H Kao, Jinsong Leng, and H Jerry Qi. 2013. "Design Considerations for Shape Memory Polymer Composites with Magnetic Particles." *Journal of Composite Materials* 47 (1): 51–63. https://doi.org/10.1177/0021998312447647.

Yu, Kai, Tao Xie, Jinsong Leng, Yifu Ding, and H. Jerry Qi. 2012. "Mechanisms of Multi-Shape Memory Effects and Associated Energy Release in Shape Memory Polymers." *Soft Matter* 8 (20): 5687–95. https://doi.org/10.1039/C2SM25292A.

Zeng, Hao, Jinsong Leng, Jianping Gu, and Huiyu Sun. 2019. "Modeling the Thermomechanical Behaviors of Shape Memory Polymers and Their Nanocomposites by a Network Transition Theory." *Smart Materials and Structures* 28 (6): 065018. https://doi.org/10.1088/1361-665X/ab1156.

Zeng, Hao, Jinsong Leng, Jianping Gu, Chenxi Yin, and Huiyu Sun. 2018. "Modeling the Strain Rate-, Hold Time-, and Temperature-Dependent Cyclic Behaviors of Amorphous Shape Memory Polymers." *Smart Materials and Structures* 27 (7): 075050. https://doi.org/10.1088/1361-665X/aaca50.

Zeng, Hao, Zhimin Xie, Jianping Gu, and Huiyu Sun. 2018. "A 1D Thermomechanical Network Transition Constitutive Model Coupled with Multiple Structural Relaxation for Shape Memory Polymers." *Smart Materials and Structures* 27 (3): 035024. https://doi.org/10.1088/1361-665X/aaae29.

Zhang, Jiamei, Guansuo Dui, and Xiaoyan Liang. 2018. "Revisiting the Micro-Buckling of Carbon Fibers in Elastic Memory Composite Plates under Pure Bending." *International Journal of Mechanical Sciences* 136 (February): 339–48. https://doi.org/10.1016/j.ijmecsci.2017.12.018.

Zhao, Qian, H. Jerry Qi, and Tao Xie. 2015. "Recent Progress in Shape Memory Polymer: New Behavior, Enabling Materials, and Mechanistic Understanding." Progress in Polymer Science, *Self-Healing Polymers* 49–50 (October): 79–120. https://doi.org/10.1016/j.progpolymsci.2015.04.001.

Zhou, Bo, YanJu Liu, and JinSong Leng. 2010. "A Macro-Mechanical Constitutive Model for Shape Memory Polymer." *Science China Physics, Mechanics and Astronomy* 53 (12): 2266–73. https://doi.org/10.1007/s11433-010-4163-2.

5 Finite Element Analysis

5.1 BACKGROUND OF FINITE ELEMENT ANALYSIS OF COMPOSITES

Any structural engineering problem can be solved in two ways: either by analytical or by numerical one. Although analytical solutions found to be accurate, they are for simple boundary conditions, loading, and linear problems. Whenever problems appear in the form of complex geometry, irregular boundary conditions, or nonlinearity (material or geometry), the analytical solution becomes ineffective. Therefore, numerical methods are developed by the researchers for obtaining such solutions.

Just consider 'n' number of partial differential equations (PDE), then we have to solve them by using a numerical method called the finite element method (FEM). Normally, whenever there is a PDE, it is difficult to solve and find the result. So what FEM does is that it converts PDE into algebraic equations, so systems of algebraic equations are formed that can be easily solved by traditional linear algebraic techniques (Tiwari and Shaikh 2019, 2021). In general terms, FEM is a method for dividing up a very complicated problem into small elements that can be solved in relation to each other. Initially, FEM was started with the application for aircraft structures, but with the wider nature of its application, researchers have started applying it to very complex boundary value problems, where complex systems which has to be solved. In those types of problems, applying FEM has become easier.

Hrennikoff (1941) and Richard Courant (1943) were the people who started work on FEM. The framework technique was developed by Hrenikoff, who depicted a planar elastic material as a group of bars and beams. Digital computers made it feasible to solve a significant number of modeling equations in the 1950s. The phrase "finite element method" was first used in a report written by Ray W. Clough in 1960. The first conference meeting on "Finite Elements" took place in 1965. In 1967, the first text on the "Finite Element Technique" was released by Zienkiewicz and Chung. The most popular FEM software programs, including ABAQUS, ANSYS, and NASTRAN, were developed in the late 1970s. CAD systems rapidly expand as a result of interactive FEM programs running on powerful computers. In the 1980s, using finite element software, algorithms were created for electric use, hydraulic movement, and temperature analysis. Using finite element analysis (FEM) and other techniques, engineers were able to assess vibration-control strategies and expand the use of bendable, movable structures in space in the 1990s. In this decade, trends have been noted that embrace completely linked solutions to hydraulic flow, structural interactions, and associated bio-mechanics issues with a higher degree of precision. Today, structural behavior can be accurately understood thanks to the creation of FEM and enormous advances in processing speed and comfort. In reality, before the invention of computers, this was unimaginable. One-dimensional components were used by Hrennikoff in 1941 to solve an elasticity issue, and McHenry followed suit in 1943. In order to represent the entire area, Courant introduced form functions over triangle subregions in 1943. Levy created the force (flexibility) technique for structural analysis in 1947, and in 1953 he created the rigidity method for structural issues.

In the Boeing Structural Dynamics Unit, Ray Clough collaborated with John Turner to determine the impact factors for bending and twisting elasticity in wings with a low aspect ratio. The process for creating the continuous tension triangle was developed by them. It was shown that a rectangle membrane element could be used instead of the continuous strain triangle membrane element to prevent shear locking by taking advantage of equilibrium stress patterns. The straight rigidity technique was used to create the node equilibrium formulas. Clough created a matrix algebra system that helps answer larger systems through the use of submatrix methods and tape storage to aid pupils learning to program who require assistance in handling issues involving finite elements. Graduate student Ari Adini, under Clough's supervision, used matrix algebra tools to solve a series of planar stress issues with triangle elements. For the duration of it all, Clough was there.

With the help of the plane-stress element and the initial location of nodes and their numbering, Clough (1960) generated a fully functional FEM program, which in the coming years was found helpful for researchers and engineers to solve arbitrary shape problems utilizing different materials. In 1965, the "Finite Element Method" came to be recognized as the "Direct Stiffness Method's" new nomenclature. Finite element rigidity matrices were first introduced in an isoparametric version by Bruce Irons and Olgierd Zienkiewicz in 1968. This discovery had a big effect on the study of finite elements. Shear locking was an issue that Wilson et al. (1973) fixed by using an eight-node solid element and a four-node planar element. Utilizing conflicting shift patterns and decreased integration, they were able to effectively fix the issue. 1991 also saw the effective completion of the implementation of model hierarchies in FEA software, which allowed FEA researchers to distinguish between discretization and idealization mistakes for the first time in the field's history. This was crucial for verification, validation, and the measurement of uncertainty.

Since the 1970s, composites have been used in a wide range of applications thanks to the introduction of novel fibers such as carbon, graphite, aramids, kevlar, and boron and the creation of new composite systems with frameworks made of metals and ceramics. Engineering and quantitative physics issues can be numerically solved using the finite element technique. Structural analysis, heat transfer, mass transit, electro-mechanical systems, and fluid movement are just some of the common issue domains in engineering and physics that can be tackled with the FEM. When a mathematical answer cannot be evaluated due to factors such as critical forms, dimensions, material properties, loadings, and boundary conditions, resolving a system of conventional or PDE is often required. Therefore, it is important to focus on a computational answer, such as the finite element technique, in order to achieve a reliable and satisfactory result. The solution of differential equations is avoided in favor of solving a set of simultaneous algebraic equations produced by the FEM model of the issue. These computational solutions produce approximations of the unknowns at a finite number of locations along the spectrum. To represent a body domain, one must divide it into a set of smaller bodies or units linked at points common to two or more elements (nodes) and border lines or surfaces; this process is known as discretization.

5.2   BENDING ANALYSIS OF LAMINATES

Application of FEM with shear deformation beam theory for flexural/bending analysis (1D) is discussed in this section. For the purpose of clarity in outlining the next actions, we shall focus only on the structural issues. For solving the one-dimensional composite beam problem with shear deformable beam theory utilizing FEM, steps are followed are indicated in Figure 5.1.

The major steps in the Finite Element Method are:

1. Discretization of the real continuum or structure – (Establishing the FE mesh)
   a. Establish the FE mesh with set coordinates, element numbers, and node numbers.
   b. The discretized FE model must be situated in a coordinate system.
   c. Elements and nodes in the discretized FE model need to be identified by "element numbers" and "nodal numbers."
   d. Nodes are identified by the assigned node numbers and their corresponding coordinates.
2. Identify the primary unknown quantity
   a. Primary unknown quantity – The first and principal unknown quantity to be obtained by the FEM.
   b. E.g.: Stress analysis: Displacement $\{u\}$ at nodes.
   c. In stress analysis, the primary unknowns are nodal displacements, but secondary unknown quantities include strains in elements, which can be obtained by the "strain–displacement relations," and unknown stresses in the elements by the stress–strain relations (the Hooke's law).
3. Interpolation functions and the derivation of Interpolation functions
   a. The interpolation function is called the "shape function" in some literature.
   b. There are different forms of interpolation functions used in FEM. The elements using the linear interpolation functions are called "Simplex elements". They are the simplest form and the most commonly used in FE formulation.
4. Derivation of the element equation
   a. The element equation relates the induced primary unknown quantity in the analysis to the action. Eg. In a structural stress analysis, Force $\{F\}$ is the action, Displacement $\{u\}$ at nodes is the primary unknown, and Stresses $\{\sigma\}$ & Strains $\{\varepsilon\}$ are secondary unknown.

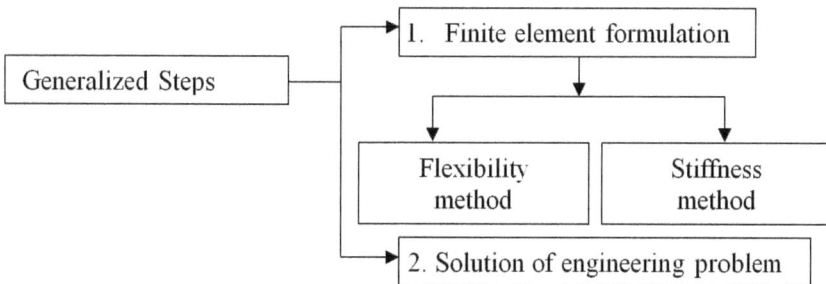

FIGURE 5.1   Steps for FEA of 1D composite beam.

b. These are generally the two methods used to derive the element equations:
   i. The Rayleigh–Ritz method and
   ii. The Galerkin method
5. Derive the overall Stiffness equation
   a. This step assembles all the individual element equations derived in Step 4 to provide the "Stiffness equations" for the entire medium.
   b. Mathematically, this equation has the form, $[K]\{q\} = \{F\}$ where $\{F\}$ is the overall stiffness matrix.
6. Solve for primary unknowns
   a. Use the inverse matrix method to solve the primary unknown quantities $\{q\}$ at all the nodes from the overall stiffness equations $\{q\} = [K]-1\{F\}$
   b. Otherwise, use the Gaussian elimination method or its derivatives to solve the nodal quantities $\{q\}$ from the equation: $[K]\{q\} = \{F\}$

STEP 1: Element selection

In the first step, a two-nodded line element/bar element is selected as an element type, as shown in Figure 5.2. The following displacement function is selected for the numerical analysis.

STEP 2: Selection of the displacement field model

Based on higher-order shear deformation theory and C^1 continuity, the displacement model components are assessed for an arbitrary shape memory polymer composites (SMPC)/shape memory polymer hybrid composite (SMPHC) one-dimensional beam with a changed displacement scheme in the 'x' and 'z' directions via the beam. This method has been used in Eq. 5.1, which lessens computation-related issues (Lal and Markad 2021; Markad and Lal, 2021).

$$\bar{u}(x,z) = u + f_1(z)\Psi_x + f_2(z)\varnothing_x; \quad \bar{w}(x,z) = w \qquad (5.1)$$

Here,

U = midplane axial displacement

w = midplane transverse displacement

Ψ_x = Normal to midplane rotation along the y-axis

$\varnothing_x$ = Slope along the x-axis $\left(\dfrac{\partial w}{\partial x}\right)$

$f_1 = C_1 Z - C_2 \alpha^3$; with $C_1 = 1$ and $C_2 = \dfrac{4}{3h^2}$

$f_2 = -C_2 \alpha^3$

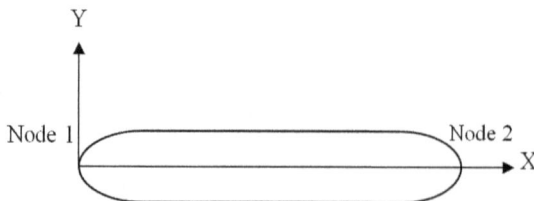

FIGURE 5.2 Element type selection.

Iso-parametric FEA has been adopted in an appropriate C^1 continuity to apply this theory with four degrees of freedom (DOFs) at each node (u, w, $\varnothing_x$, Ψ_x). Degree of freedom is nothing but the number of independent variables required to specify the motion of a body completely. This approach has been adopted in the formulation to reduce cumbersome computations without affecting the accuracy of the solution.

Displacement vector for the modified C^1 continuous model can be written as Eq. 5.2.

$$\{q\} = [u \quad w \quad \varnothing_x \quad \Psi_x]^T \tag{5.2}$$

STEP 3: Establish the relationship between stress, strain, and displacement
This type of relationship derivation is important for each FEA. Assume the relation stated in Eq. 5.3 for strain and displacement is valid,

$$\varepsilon_x(x, y) = \frac{\partial u}{\partial_-} \tag{5.3}$$

5.2.1 STRESS–STRAIN–DISPLACEMENT BEHAVIOR

The total strain vector consists of linear strain (in terms of midplane deformation and rotation of normal and higher-order terms), nonlinear strain (Von-Karman type) associated with the displacement for fiber reinforced polymer composites, which can be expressed as Eq. 5.4:

$$\{\bar{\varepsilon}\} = \{\bar{\varepsilon}^L\} + \{\bar{\varepsilon}^{NL}\} \tag{5.3}$$

According to Green–Lagrange theory, differentiating Eq. 5.4 after substituting all essential values with respect to x

$$\varepsilon_x = \frac{\partial u_0}{\partial x} + zC_1\frac{\partial \Psi_x}{\partial x} - z^3\frac{4}{3h^2}\left(\frac{\partial \Psi_x}{\partial x} + \frac{\partial \phi_x}{\partial x}\right) \tag{5.5}$$

Rearranging the terms,

$$\varepsilon_x = \frac{\partial u_0}{\partial x} + z\left(C_1\frac{\partial \Psi_x}{\partial x} - \frac{4z^2}{3h^2}\left\{\frac{\partial \Psi_x}{\partial x} + \frac{\partial \phi_x}{\partial x}\right\}\right) \tag{5.6}$$

$$\varepsilon_1 = \varepsilon_1^0 + z\left(k_1^0 + z^2k_1^1\right) \tag{5.7}$$

Where, $k_1^0 = C_1\dfrac{\partial \Psi_x}{\partial x}$; $k_1^1 = -C_2\left\{\dfrac{\partial \Psi_x}{\partial x} + \dfrac{\partial \phi_x}{\partial x}\right\}$

Acquiring the same differentiation for strain, total linear and nonlinear strain are obtained as $\{\varepsilon\} = \{\varepsilon_L\}$ and $\{\varepsilon\} = \{\varepsilon_{NL}\}$.

Assuming that the strains are much smaller than the rotations (in the von-Karman sense), one can rewrite the nonlinear strain vector $\{\varepsilon_{nl}\}$ as Eq. 5.8.

$$\{\varepsilon_{NL}\} = \frac{1}{2}[A]\{\varnothing\} \tag{5.8}$$

where $[A] = \dfrac{1}{2}\begin{bmatrix} w_{,x} & 0 & w_{,y} & 0 & 0 \\ 0 & w_{,x} & w_{,y} & 0 & 0 \end{bmatrix}^T$ and $\phi = \dfrac{1}{2}\begin{bmatrix} w_{,x} \\ w_{,y} \end{bmatrix}$,

where (,) denotes partial differential.

$$\begin{Bmatrix} \varepsilon_1^0 \\ \varepsilon_2^0 \\ \varepsilon_3^0 \\ k_1^0 \\ k_3^0 \\ k_1^1 \\ k_3^1 \end{Bmatrix} = \begin{bmatrix} \dfrac{\partial}{\partial x} & 0 & 0 & 0 \\ 0 & \dfrac{\partial}{\partial z} & 0 & 0 \\ \dfrac{\partial}{\partial z} & \dfrac{\partial}{\partial x} & 0 & 0 \\ 0 & 0 & C_1\dfrac{\partial}{\partial x} & 0 \\ 0 & 0 & C_1\dfrac{\partial}{\partial z} & 0 \\ 0 & 0 & -C_2\dfrac{\partial}{\partial x} & -C_2\dfrac{\partial}{\partial x} \\ 0 & 0 & -C_2\dfrac{\partial}{\partial z} & -C_2\dfrac{\partial}{\partial z} \end{bmatrix} \begin{Bmatrix} u \\ w \\ \Psi_x \\ \phi_x \end{Bmatrix} \tag{5.9}$$

where $[B] = \begin{bmatrix} \dfrac{\partial}{\partial x} & 0 & 0 & 0 \\ 0 & \dfrac{\partial}{\partial z} & 0 & 0 \\ \dfrac{\partial}{\partial z} & \dfrac{\partial}{\partial x} & 0 & 0 \\ 0 & 0 & C_1\dfrac{\partial}{\partial x} & 0 \\ 0 & 0 & C_1\dfrac{\partial}{\partial z} & 0 \\ 0 & 0 & -C_2\dfrac{\partial}{\partial x} & -C_2\dfrac{\partial}{\partial x} \\ 0 & 0 & -C_2\dfrac{\partial}{\partial z} & -C_2\dfrac{\partial}{\partial z} \end{bmatrix}$ has order $[7 \times 4]$.

STEP 4: Deriving element stiffness matrix

The idea of stiffness influence coefficients was originally used to build element stiffness matrices and element equations; however, this requires some familiarity with structural analysis. Stiffness matrix is written as Eq. 5.10:

$$[K] = [B]^T [D][B] \tag{5.10}$$

In this expression, the order of $[B]^T$ is 4×7, $[D]$ is 7×7, and $[B]$ is 7×4. Here $[D]$ is elastic stiffness matrix.

For the plane-stress scenario utilizing the thermo-elastic constitutive model, the relationship between stress and strain may be written as Eq. 5.11 (Lal and Markad 2018; Markad et al. 2022).

$$\left\{ \begin{matrix} \overline{\sigma}_x \\ \overline{\tau}_{xz} \end{matrix} \right\} = \begin{bmatrix} \overline{\mathbb{Q}}_{11} & 0 \\ 0 & \overline{\mathbb{Q}}_{55} \end{bmatrix} \left\{ \{\overline{\varepsilon}^L\} + \{\overline{\varepsilon}^{NL}\} - \{\overline{\varepsilon}^T\} \right\} \tag{5.11}$$

where $\overline{\mathbb{Q}}_{11} = \mathbb{Q}_{11} \sin^4 \theta_k + 2(\mathbb{Q}_{12} + 2\mathbb{Q}_{66}) \cos^2 \theta_k \cos^2 \theta_k$ and $\overline{\mathbb{Q}}_{55} = \mathbb{Q}_{55} \cos^2 \theta_k +$ $\mathbb{Q}_{44} \sin^2 \theta_k$ with $\mathbb{Q}_{11} = \dfrac{E_{C1}}{1 - \mu_{12}\mu_{21}}$; $\mathbb{Q}_{22} = \dfrac{E_{C12}}{1 - \mu_{12}\mu_{21}}$; $\mathbb{Q}_{12} = \dfrac{\mu_{C12}E_{C2}}{1 - \mu_{12}\mu_{21}}$; $\mu_{21} = \dfrac{\mu_{C12}E_{C2}}{E_1}$; $\mathbb{Q}_{55} = G_{C12}$ and $\mathbb{Q}_{44} = G_{C13}$.

Here θ_k = fiber alignment; E_{C1} = longitudinal modulus; E_{C2} = transverse modulus; G is the shear modulus; and μ is Poisson's ratios.

5.2.2 STRAIN ENERGY OF BEAM

By conservation of mechanical energy principle, the external work done due to applied force P is transferred into the internal strain energy U of beam. The SMPC beam's strain energy (Π_1) after being significantly deformed may be represented as Eq. 5.12:

$$\Pi_1 = U_L + U_{NL} \tag{5.12}$$

The fiber-reinforced polymer composite beam's linear stain energy (U_L) is expressed as Eq. 5.13.

$$U_L = \iint \frac{1}{2} \{\overline{\varepsilon}^L\}^T [D] \{\overline{\varepsilon}^L\} dA \tag{5.13}$$

where $[D]$ and $\{\overline{\varepsilon}^L\}$ are elastic stiffness matrix and linear strain vector, respectively.

The fiber-reinforced polymer composite beam's nonlinear stain energy (U_{NL}) is expressed as Eq. 5.14.

$$U_{NL} = \iint \frac{1}{2} \{\overline{\varepsilon}^L\}^T [D_1] \{\overline{\varepsilon}^{NL}\} + \frac{1}{2} \{\overline{\varepsilon}^{NL}\}^T [D_2] \{\overline{\varepsilon}^L\} + \frac{1}{2} \{\overline{\varepsilon}^{NL}\}^T [D_3] \{\overline{\varepsilon}^{NL}\} dA \tag{5.14}$$

Work done due to the externally applied uniformly distributed load can be written as Eq. 5.15

$$\Pi_3 = W_q = \int\int P(x,z)wdA \tag{5.15}$$

where, $P(x,z)$ is evenly distributed load.

STEP 5: Assembling all elemental equations and formation of the global stiffness matrix

5.2.3 STRAIN ENERGY OF THE BEAM ELEMENT

In the present study, a C^1 one-dimensional Hermitian beam element with four DOFs per node is employed. Expressed as this type of beam element geometry and the displacement vector for this element in Eq. 5.16:

$$\{q\} = \sum_{i=1}^{NN} N_i\{q\}_i ; \quad x = \sum_{i=1}^{NN} N_i x_i \tag{5.16}$$

where

N_i – interpolation function for the ith node
$\{q\}_i$ – vector of unknown displacements for i^{th} node
NN – the number of nodes per element
x_i – Cartesian coordinate

Linear interpolation for axial displacement and rotation of the normal and Hermite cubic interpolation functions for transverse displacement and slope are chosen. Eq. 5.12 may be represented as, using FEM, can be expressed as Eq. 5.17,

$$\Pi_1 = \sum_{e=1}^{NE} \Pi_a^{(e)} = \sum_{e=1}^{NE} \left(U_L^{(e)} + U_{NL}^{(e)} \right) \tag{5.17}$$

where NE and (e) denote the number of elements and elemental, respectively.
Eq. 5.17 can be further evaluated as Eq. 5.18,

$$\Pi_1 = \frac{1}{2}\sum_{e=1}^{NE} \left[\{q\}^{T(e)} \left[K_l + K_{nl} \right]^e \{q\}^{(e)} \right] = \{q\}^T \left[K_l + K_{nl} \right]\{q\} \tag{5.18}$$

Here, $[K_{nl}] = \frac{1}{2}[K_{nl1}] + [K_{nl2}] + \frac{1}{2}[K_{nl3}]$ and $[K_l], [K_{lth}], [K_{1nl}], [K_{2nl}], [K_{3nl}], \{q\}$ are defined as global linear and nonlinear stiffness matrices of plate and thermal effect and global displacement vector, respectively, which are stated as Eqs. 5.19 and 5.20.

$$K_l^{(e)} = \frac{1}{2}\int_A [B]^{(e)^T} D[B]^{(e)} dA, \quad K_{nl}^{(e)} = \frac{1}{2}\int_A [B]^{(e)^T} D_l[B]^{(e)} dA \tag{5.19}$$

$$K_{nl1}^{(e)} = \frac{1}{2}\int_A [B]^{(e)^T} D_2[B]^{(e)} dA, \quad K_{nl2}^{(e)} = \frac{1}{2}\int_A [B]^{(e)^T} D_3[B]^{(e)} dA \tag{5.20}$$

where

$$D = \int_{-\frac{k}{2}}^{\frac{k}{2}} [T]^T [\bar{Q}][T] dz = \begin{bmatrix} A_{11} & B_{11} & E_{11} & 0 & 0 \\ B_{11} & D_{11} & F_{11} & 0 & 0 \\ E_{11} & F_{11} & H_{11} & 0 & 0 \\ 0 & 0 & 0 & A_{55} & D_{55} \\ 0 & 0 & 0 & D_{55} & F_{55} \end{bmatrix},$$

$$D_1 = \begin{bmatrix} A_{11} & 0 \\ B_{11} & 0 \\ E_{11} & 0 \\ 0 & A_{55} \\ 0 & D_{55} \end{bmatrix}, D_3 = \begin{bmatrix} A_{11} & 0 \\ 0 & A_{55} \end{bmatrix},$$

$$D_2 = \begin{bmatrix} A_{11} & B_{11} & E_{11} & 0 & 0 \\ 0 & 0 & 0 & A_{55} & D_{55} \end{bmatrix}$$

with $(A_{11}, B_{11}, D_{11}, E_{11}, F_{11}, H_{11}) = \int_{-k/2}^{k/2} Q_{11}(1, z, z^2, z^3, z^4, z^6) dz$.

Using FEM, Eq. 5.15 can be modified as Eq. 5.21,

$$\Pi_3 = \sum_{e=1}^{NE} \Pi_3^{(e)} = \sum_{e=1}^{NE} \{q\} \tag{5.21}$$

STEP 6: Solve for primary unknowns

5.2.4 Governing Equation of Bending

The governing equation of the nonlinear static analysis can be derived using the variational principle, which is a generalized form of the principle of virtual displacement. For the bending analysis, the minimization of the first variation of total potential energy Π $(\Pi_1 + \Pi_f - \Pi_2)$ with respect to the displacement vector is given by Eq. 5.22,

$$\partial(\Pi_1 + \Pi_f - \Pi_2) = 0 \tag{5.22}$$

By substituting the previous correlations in Eq. 5.22, the resultant equation is obtained as Eq. 5.23,

$$[K_l + K_{nl}]\{q\} = \{F\} \tag{5.23}$$

Stiffness matrix $[K]$ consists of linear and nonlinear plate and foundation stiffness matrices. Parameters $\{q\}$ and $\{F\}$ are transverse deflection and force vectors, respectively.

The solution of Eq. 5.23 can be obtained using standard solution procedures such as direct iterative, incremental, and/or Newton–Raphson methods, etc. However, the Newton–Raphson method is one of the most popular and widely used solution procedures due to its fast convergence at higher amplitudes.

5.3   CASE STUDY: FLEXURAL ANALYSIS OF A CNT-REINFORCED COMPOSITE BEAM

Consider the PMMA matrix reinforced with single-walled carbon nanotube (SWCNT) initially, further replace it with multi-walled carbon nanotube (MWCNT), and find effective material properties. Also under the application of a uniformly distributed load, state the performance of the carbon nanotube-reinforced composite (CNTRC) beam in terms of transverse central deflection. This case study deals with the two-phase composite material structural analysis validation and program convergence only.

A uniformly distributed load is applied over the composite beam shown by q_0 and W_0 is the transverse central deflection occurring along the z direction. A user-interactive computer program in the MATLAB [R2015a] environment has been developed to evaluate the numerical results. The detailed numerical solution for bending analysis is presented here for MWCNT reinforced composite beam panels. PMMA is selected for matrix, $E = 3.52$ Gpa, $\rho_m = 1150\frac{Kg}{m^3}$, $\upsilon_m = 0.34$.

Figure 5.3 shows the convergence study for the different number of elements (*nel*) clamped supported MWCNTRC, which is the combination of matrix and MWCNT/SWCNT. As *nel* increases, the nondimensional transverse central deflection (NTCD) in the SWCNTRC and MWCNTRC beams increases and converges from *nel* equal

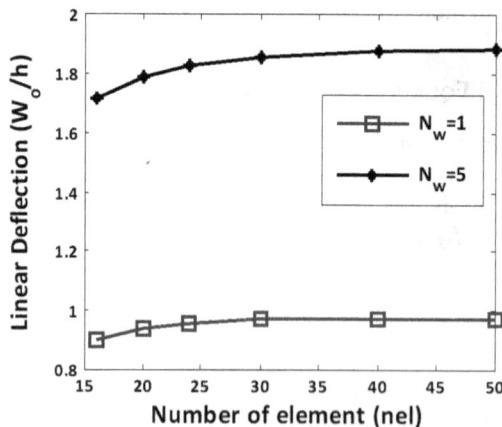

FIGURE 5.3   Convergence near to exact solution for the deflection in composite beam with variation in number of elements.

to 25 and onward. Hence for further study, a total of 30 elements are considered in the present analysis.

The beam's deflection as a function of applied force must be calculated in order to solve the nonlinear bending issue. In this scenario, we will suppose that the transverse static load is homogeneous, with $Q(X, t) = q = q_0$. Therefore, W does not rely on the passage of time. In this subsection, we give numerical findings for the flexural properties of CNTRC beams supported by elastic bases and situated in different temperature settings. The material characteristics of CNTRCs must be established initially. Matrix poly (methyl methacrylate), or PMMA, and CNT material properties are assumed as,

$$\rho_m = 1150\frac{Kg}{m^3}, \ \upsilon_m = 0.34, \ \alpha_m = 45(1 + 0.0005\Delta T) \times \frac{10^{-6}}{K}, \ E_m = 3.52 - 0.0034T \ \text{GPa}.$$

During the analysis, room temperature was taken as 300 K, and T = room temperature + ΔT. Single-walled CNT taken for reinforcement in matrix, and $(m, n) = (10, 10)$ chirality SWCNT $E_{cnt_1} = 600\,\text{GPa}$, $E_{cnt_2} = 10\,\text{GPa}$, $G_{cnt_{12}} = 17.2$ GPa, $\upsilon_{cnt_{12}} = 0.19$. Comparison of load-central deflection for the different volume fractions under uniformly distributed loads is indicated in Figure 5.4.

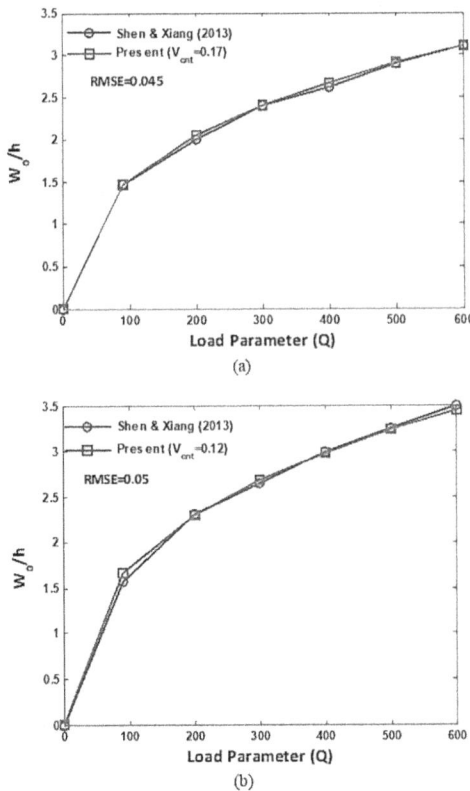

FIGURE 5.4 Comparison of load-central deflection for (a) $V_{cnt} = 0.17$ and (b) $V_{cnt} = 0.12$ under uniformly distributed load.

5.4   CASE STUDY: FLEXURAL ANALYSIS OF SHAPE MEMORY
POLYMER HYBRID COMPOSITES (SMPHC)

Here the performance of the SMPC against the dynamic temperature variation is observed under uniformly distributed load (UDL), but instead of this two-phase composite material, an MWCNT SMPHC beam is utilized to see how it responds. To observe the performance of this SMPHC beam against dynamic temperature variation and loading, the following results are necessary to study:

Figure 5.5 shows the effect of SWCNT addition to the shape memory polymer matrix over the transverse central deflection of the SMPHC beam. For the analysis, beam dimensions are taken as $a/h = 15$, and $h = 0.33$ m, and they are the same in all subsequent results unless otherwise mentioned, with a ply orientation of (0/90/90/0) under clamped conditions and UDL. The result clearly revealed that with the addition of SWCNT to the present utilized model, the dimensionless central deflection is significantly reduced because of the improvement in strength of the modified matrix, interfacial bonding with fiber and hence SMPC.

Table 5.1 clearly differentiates between the performance of SMPC and SMPHC beams with the utilization of SWCNT and MWCNT with different lamination or stacking sequences of layups. In the table, the third column shows the transverse central deflection (W_0/h) of the SMPC beam having an aspect ratio of 15 under clamped boundary conditions at 328 K and constant unit UDL. There are certain important observations that come out of the study. First, with the addition of SWCNT into SMP with a V_{cnt} of 0.5%, the overall central deflection of the composite beam is reduced by 19%, 17.6%, and 6% for two-, four-, and six-layered composite beams as compared to SMPC. Also, instead of SWCNT, if MWCNT is utilized, with $N_w = 2$, then dimensionless central beam deflection decreases by 9%, 8.26%, and 2.5%, and with $N_w = 3$, it decreases by 5.84%, 5.5%, and 1.5%, respectively, as compared to SMPC.

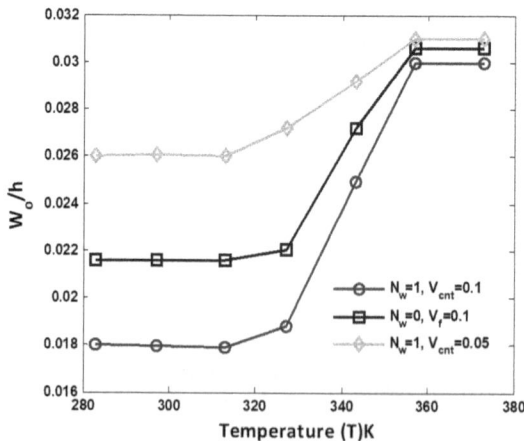

FIGURE 5.5   Effect of SWCNT over transverse central deflection of the SMPC beam.

TABLE 5.1
Effect of the Number of Walls of CNT and Lamination Scheme Over Central Deflection of SMPC Beam

Number of Walls of CNT	Lamination Scheme	TCD of two-phase SMPC (N_w=0)	TCD of three-phase Composite
$N_w = 1$	0/90	0.02807	0.0227
	0/90/90/0	0.026794	0.02207
	45/0/0/0/0/45	0.018741	0.01762
$N_w = 2$	0/90	–	0.02554
	0/90/90/0	–	0.02458
	45/0/0/0/0/45	–	0.01828
$N_w = 3$	0/90	–	0.02643
	0/90/90/0	–	0.02532
	45/0/0/0/0/45	–	0.01846

MATLAB CODE: 5 Flexural Analysis of a Multi-Walled CNTRC Beam

```
clc
clear all
%% Beam material property Evaluation
rhom=1150;
Mum=0.34;
T=300;
T0=300;
deltaT=T-T0;
% T=T0+deltaT;
Em=(3.52-0.0034*T)*10^9;   % Temperature dependent property of
matrix
E11cn=600*10^9;
E22cn=10*10^9;
G12cn=17.2*10^9;
Mucn=0.19;
Vcn=0.17;
Vm=1-Vcn;
alpham=45*(1+0.0005*deltaT)*10^(-6);
% alpham=45*1e-9;
alpha11cn=3.4854*1e-6;
alpha12cn=5.1682*1e-6;
%%%%%%%%%%%%%%%%%%%%%%%%%% CNT volume fraction
% Vcn=0.12
%%%%%%%%%%%%%%%%%%%%%%%%%%
% Q1=0.137;
% Q2=1.022;
% Q3=1.715;
%%%%%%%%%%%%%%%%%%%%%%%%%%%
% for Vcn=0.17
%%%%%%%%%%%%%%%%%%%%%%%%%%%%%%
```

```
Q1=0.142;
Q2=1.626;
Q3=1.138;
%%%%%%%%%%%%%%%%%%%%%%%%%%%%%%%%%
% Vcn=0.28
%%%%%%%%%%%%%%%%%%%%%%%%%%%%%%%%%
% Q1=0.141;
% Q2=1.585;
% Q3=1.109;
%%%%%%%%%%%%%%%%%%%%%%%%%%%%%%%%%

%% Effective material property calculation:

E1=Q1*E11cn*Vcn+Em*(1-Vcn);
E2=Q2*E22cn*Em/((1-Vcn)*E22cn+Vcn*Em);
Gm=Em/(2*(1+Mum));
G12=Q3*G12cn*Gm/((G12cn*Vm+Vcn*Gm));
V12=Mucn*Vcn+Mum*(1-Vcn);
alp1=(Vcn*E11cn*alpha11cn+(1-Vcn)*Em*alpham)/
(Vcn*E11cn+(1-Vcn)*Em);
alp2=(1+Mucn)*Vcn*alpha12cn+(1+Mum)*(1-Vcn)*alpham-(V12*alp1);
G13=G12;
 G23=0;
V21=V12*(E2/E1);



a=4.54;
b=1.01;  %input('Width of beam=');
r=20;%input('Thickness ratio a/h =');
h=a/r;
nL=1;
h1=h/nL;
 I=(b*h^3)/12;
    k1=000;      %input('Non dimensional Linear Winkler
foundation stiffness=');
    k2=00;       %input('Non dimensional Pasternak (shear)
foundation stiffness=');
    k3=0;        %input('Non dimensional nonlinear foundation
stiffness=');
%
thetak=zeros(1,nL);
for k =1:nL
      thetak(1,1)=0;        %input('stacking sequence
thetak(1,k)=');
%        thetak(1,2)=pi/2;
%        thetak(1,3)=0;
end
h=a/r;
hf=h;
nel=30;%input(' number of elements=');
```

```
nnel=2;%input('number of nodes per element=')
tleng=a;
leng=tleng/nel; % element length of equal size
s=leng;
% area=b*h; % cross-sectional area of the beam
ndof_q=4; % degree of freedom per node
L=(nnel-1)*(nel)+1;  % total no of node in one side
nnode=L;                      % total number of nodes in system
edof_q=nnel*ndof_q;           %degree of freedom per element
sdof_q=nnode*ndof_q;          % degree of freedom of system
nglx =3;ngly =3;                      % 3*3 integration rule
% Wmax=1.5;%input('amplitude ratio=');
%------------------------------------------------------------------------
% input data for nodal coordinate values
% gcoord(i,j) where i-node no. and j- x or y
%------------------------------------------------------------------------
% a4=(a)/(2*(nel)^.5);
% b4=(b)/(2*(nel)^.5);
a4=(a)/(L-1);
b4=(b)/(L-1);
% a4=(a)/(4*(nel)^1);
% b4=(b)/(4*(nel)^1);

% loop for putting y-coordinate values of nodes of system
for k=1:L
for p=((k-1)*L+1):(k*L)
     gycoord(p,2)=b4*(k-1);
end
end
% loop for putting x coordinate values of nodes of sysyem
for k=1:L
for p=k:L
     gxcoord(p,1)=a4*(k-1);
end
end
%------------------------------------------------------------------------
% input data for nodal connectivity for each element
% nodes(i,j) where i-element no. and j-connected nodes from 1
to 9, for nine-node element
%-----------------------------------------------------------------%
a0=nel^1/1;
% a0=1+1;
for k=1:nel
 for p=k:a0
%    if i>(k-1)*a0 & i<=k*a0
%        L=1*(nel)^1+1;
%        a1=(k-1)*a0;
%....four node..........
  a2=1*(k);
  nodes(p,1)=1*a2;
  nodes(p,2)=1*a2+1;
```

```
end
end%
%-----------------------------------------------------------------------------
mnn=1;      %input('mode=');
COC1=0.00;   %input('coefficient of correlation for Ec=');
COC2=0.00;   %input('coefficient of correlation for Vc=');
COC3=0.00;   %input('coefficient of correlation for Em=');
COC4=0.00;   %input('coefficient of correlation for Vm=');
COC5=0.00;   %input('coefficient of correlation for n=');
COC6=0.00;   %input('coefficient of correlation for Ep=');
COC7=0.00;   %input('coefficient of correlation for Vp=');
COC8=0.00;   %input('coefficient of correlation for alphap=');
COC9=0.10;   %input('coefficient of correlation for alphac=');
COC10=0.00;  %input('coefficient of correlation for alpham=');
COC11=0.00;  %input('coefficient of correlation for kt=');
COC12=0.00;  %input('coefficient of correlation for kb=');
COC13=0.00;  %input('coefficient of correlation for qm

en=0;
enu=0;
n=1;%input('volume fraction index for n=');


 value_wmax=[0 1];
 for zz=1:2
 Wmax=value_wmax(1,zz);

value_q=[0.001 90 200 300 400 500 600];
for pp=1:7
    qm=value_q(1,pp);
%----------------------------------------------------------------------------
% -----------------Simply supported beam...............
if nel==30
%    bcdof=[1 2 3 4 121 122 123 124]; % clamp-clamp
%    bcval=zeros(1,8);
%    bcdof=[1 2 3 4];    % clamp- free
%    bcval=zeros(1,4);
%    bcdof=[1 2 3 4 121 122]; % clamp-supported
%    bcval=zeros(1,6);
%     bcdof=[1   117 ]; % simply supported...*
%       bcval=zeros(1,4);
%       bcdof=[1 121 122 123 124]; % Hinged-Hinged
%       bcval=zeros(1,5);
   bcdof=[1 4 117 118 119]; % simply supported
   bcval=zeros(1,4);
%    bcdof=[1 117 118 119]; % simply supported
%    bcval=zeros(1,5);

elseif nel==24
 bcdof=[1 2 3 4 97 98 99 100];
 bcval=zeros(1,8);
```

```
elseif nel==20
 bcdof=[1 2 3 4 81 82 83 84];
 bcval=zeros(1,8);

elseif nel==16
 bcdof=[1 2 3 4 65 66 67 68]; % clamp-clamp
 bcval=zeros(1,8);
%
%    bcdof=[1 2 3 4]; % Hinged-clamp
%    bcval=zeros(1,4);
% % %
%    bcdof=[1 65 66 67 68]; % Hinged-Hinged
%    bcval=zeros(1,5);

%      bcdof=[1 2 3 4];   % clamp- free
%      bcval=zeros(1,4);
% %
%    bcdof=[1 2 65 66]; % simply supported
%    bcval=zeros(1,4);


elseif nel==10
%     bcdof=[1 2 3 4];
%     bcval=zeros(1,4);

  bcdof=[1 2 3 4 39 40 41 42];
  bcval=zeros(1,8);

%   bcdof=[1 45 46];
%   bcval=zeros(1,3);
% %
elseif nel==4
   bcdof=[1 2 3 4 15 16 17 18];
   bcval=zeros(1,8);
%
%   bcdof=[1 2 21 22];
%   bcval=zeros(1,4);

elseif nel==2
   bcdof=[1 2 3 4 9 10 11 12];
   bcval=zeros(1,8);
end


% % input data for boundary conditions
% % --------------------------------------------------------------------------------
% % ------------------Simply supported beam................ .......
% % if nel==40
% %
% %    bcdof=[1 2 3 4 161 162 163 164]; % clamp-clamp
% %    bcval=zeros(1,8);
%
```

```
% %    bcdof=[1 2 3 4 121 122 123 124]; % clamp-clamp
% %    bcval=zeros(1,8);
% %    bcdof=[1 2 3 4];   % clamp- free
% %    bcval=zeros(1,4);
% %    bcdof=[1 2 3 4 121 122]; % clamp-supported
% %    bcval=zeros(1,6);
% %    bcdof=[1 2 161 162]; % simply supported
% %    bcval=zeros(1,4);
% %    bcdof=[1 2 4 121 122 124]; % Hinged-Hinged
% %    bcval=zeros(1,6);
%  if nel==30
% %    bcdof=[1 2 3 4 121 122 123 124]; % clamp-clamp
% %    bcval=zeros(1,8);
% %    bcdof=[121 122 123 124];   % clamp- free
% %    bcval=zeros(1,4);
% %    bcdof=[1 2 3 4 121 122]; % clamp-supported
% %    bcval=zeros(1,6);

% % %
%
% elseif nel==24
%  bcdof=[1 2 3 4 97 98 99 100];
%  bcval=zeros(1,8);
%
% %  bcdof=[1 2 97 98];
% %  bcval=zeros(1,8);
%
% elseif nel==20
%  bcdof=[1 2 3 4 81 82 83 84];
%  bcval=zeros(1,8);
%
% elseif nel==16
%  bcdof=[1 2 3 4 65 66 67 68]; % clamp-clamp
%  bcval=zeros(1,8);
% %
% %    bcdof=[1 2 3 4]; % Hinged-clamp
% %    bcval=zeros(1,4);
% % % %
% %    bcdof=[1 65 66 67 68]; % Hinged-Hinged
% %    bcval=zeros(1,5);
%
% %    bcdof=[1 2 3 4];   % clamp- free
% %    bcval=zeros(1,4);
% % %
% %    bcdof=[1 2 65 66]; % simply supported
% %    bcval=zeros(1,4);
%
%
% elseif nel==10
% %    bcdof=[1 2 3 4];
% %    bcval=zeros(1,4);
```

```
%
%    bcdof=[1 2 3 4 39 40 41 42];
%    bcval=zeros(1,8);
%
% %    bcdof=[1 45 46];
% %    bcval=zeros(1,3);
% % %
% elseif nel==4
%     bcdof=[1 2 3 4 15 16 17 18];
%    bcval=zeros(1,8);
% %
% %    bcdof=[1 2 21 22];
% %    bcval=zeros(1,4);
%
% elseif nel==2
%     bcdof=[1 2 3 4 9 10 11 12];
%    bcval=zeros(1,8);
% end

% % value_qm=[-100 -200 -300  -400 -500  -600 -700  -800];
% % value_qm=[100 200 300  400 500  600 700  800];
% % for pp=1:1
%      qm=input('load=');
%%------------------------------------------------------------------------
% COC1=0.1;%input('coefficient of correlation for E1=');
% COC2=0;%input('coefficient of correlation for E2=');
% COC3=0;%input('coefficient of correlation for G12=');
% COC4=0;%input('coefficient of correlation for G13=');
% COC5=0;%input('coefficient of correlation for G23=');
% COC6=0;%input('coefficient of correlation for V12=');
% COC7=0;%input('coefficient of correlation for k1=');
% COC8=0;%input('coefficient of correlation for k2=');%%loop
for z-coordinates of various layers
% COC9=0;%input('coefficient of correlation for k3=');
% en=0;

%%%%%%%%%%%%%%%%%%%%%%%%%%%%%%%%%%%%%%%%%%%%%%%%%%%%%%%%%%%%%%%%%%%
%.... ...............................................................
%...........computation of A and its differentials---------------

  %...................................................................
.......
A=0;
%...................................................................
B=0;
%...................................................................
..
D=0;
%...................................................................
E=0;
%...................................................................
```

```
F=0;
%...................................................
H=0;
%...................................................
Df=0;

  zk=zeros(1,nL+1);
  for k=2:2
%h1=(h+2*hp)/nL;
zk(1,1)=-(h/2);
zk(1,k)=zk(1,(k-1));
zk(1,k+1)=zk(1,k)+h;
zk(1,k+2)=zk(1,k+1);
%-------------------------------------------------------------------
A=fun_QbktA(thetak,E2,E11cn,G12,V12,V21,Q1,Q3,Vcn,Vm,Em,G12cn,
Gm)*(h)+A;
% A=fun_QbktA(E11cn,G12,Q1,Vcn,Vm,Em,G12cn,Gm,h,Q3)+A;
%-------------------------------------------------------------------
%

B=fun_QbktB(thetak,E2,E11cn,G12,V12,V21,Q1,Q3,Vcn,Vm,Em,G12cn,
Gm)*(h^2)+B;
%-------------------------------------------------------------------
%

D=fun_QbktD(thetak,E2,E11cn,G12,V12,V21,Q1,Q3,Vcn,Vm,Em,G12cn,
Gm)*(h^3)+D;
% D=fun_QbktD(E11cn,G12,Q1,Vcn,Vm,Em,G12cn,Gm,h,Q3)+D;
%-------------------------------------------------------------------
%
E=fun_QbktE(thetak,E2,E11cn,G12,V12,V21,Q1,Q3,Vcn,Vm,Em,G12cn,
Gm)*(h^4)+E;
%-------------------------------------------------------------------
%
F=fun_QbktF(thetak,E2,E11cn,G12,V12,V21,Q1,Q3,Vcn,Vm,Em,G12cn,
Gm)*(h^5)+F;
% F=fun_QbktF(E11cn,Q1,Vcn,Vm,Em,G12cn,Gm,h,Q3)+F;
%-------------------------------------------------------------------
%
H=fun_QbktH(thetak,E2,E11cn,G12,V12,V21,Q1,Q3,Vcn,Vm,Em)*(h^7)+H;
%
%
A11=A(1:1,1:1);
%-------------------------------------------------------------------
z1=zeros(1,1); z2=zeros(1,1);z3=zeros(1,1);z4=zeros(1,1);
%-------------------------------------------------------------------
A22=A(5:5,5:5);
%-------------------------------------------------------------------
B11=B(1:1,1:1);
%-------------------------------------------------------------------
D11=D(1:1,1:1);
```

```
%-----------------------------------------------------------------------------
D22=D(5:5,5:5);
%-----------------------------------------------------------------------------
E11=E(1:1,1:1);
%-----------
F11=F(1:1,1:1);
%-----------
F22=F(5:5,5:5);
%------------------------------------------------------------------
H11=H(1:1,1:1);

%-----------------------------------------------------------------------------
D2=[A11 B11 E11 z1 z1;
    B11 D11 F11 z1 z1;
    E11 F11 H11 z1 z1;
    z2 z2 z2 A22 D22;
    z2 z2 z2 D22 F22]*1;
 % %.....................................................
ff_w_e=zeros(nel,30);
disp_w_e=zeros(nel,30);
%%%%%%%%%%%% Calculation for (mass matrix)
K1.....................
% K1=(k1*E2*h^3)/(b^4);
K1=(k1*a^4)/(Em*I);
dK1d1=(1*E2*h^3)/(b^4);
dK2d1=0; dK1d11=(k1*1*h^3)/(b^4);
dK2d11=0;
%------------------------------------------------------------------
%  K2=(k2*E2*h^3)/(b^2);
K2=(k2*a^2)/(Em*I);
 dK1d2=0;dK2d2=(1*E2*h^3)/(b^2);
 dK1d22=0;dK2d22=(k2*1*h^3)/(b^2);
%------------------------------------------------------------------
K3=(k3*E2*h)/(a^4);
dK3d3=(1*E2*h)/(a^4);
dK3d33=(k3*1*h)/(a^4);
%Dfnl=[K3 0 0;0 0 0;0 0 0];;
%------------------------------------------------------------------
      %final Df matix
 dKfd1=[dK1d1 0 0;0 dK2d1 0;0 0 dK2d1];
 dKfd2=[dK1d2 0 0;0 dK2d2 0;0 0 dK2d2];
 dKfd3=[dK1d11 0 0;0 dK2d22 0;0 0 dK2d22];
%------------------------------------------------------------------
 Df=[K1 0 0;
     0 K2 0;
     0 0 K2];
%------------------------------------------------------------------
Dfnl=[K3 0 0;
      0 0 0;
      0 0 0];
%------------------------------------------------------------------
```

```
 for sh=1:100
%dFF2d9=zeros(sdof_q,1);

K_q_q=zeros(sdof_q,sdof_q); % initialization of system matrix
Kf=zeros(sdof_q,sdof_q);
Kf1=zeros(sdof_q,sdof_q);
Kfs=zeros(sdof_q,sdof_q);
Kf2=zeros(sdof_q,sdof_q);
Kf3=zeros(sdof_q,sdof_q);
Kfnl3=zeros(sdof_q,sdof_q);
Kfnl33=zeros(sdof_q,sdof_q);Kfnl=zeros(sdof_q,sdof_q);
%,,,,,,,,,,,,,,,,,,random geometric,,,,,,,,,,,,,,,,,,,,,,,,,,,,,,
Ms=zeros(sdof_q,sdof_q);
FD=zeros(sdof_q,1);
ff=zeros(sdof_q,1);
dffEm=zeros(sdof_q,1);
dFF2d2=zeros(sdof_q,1);
index_q=zeros(nnel*ndof_q,1);
index_qf=zeros(nnel*ndof_q,1);
index_q3f=zeros(nnel*ndof_q,1);
index_q4df2d2=zeros(nnel*ndof_q,1); % initialization of index
vector
%....................................
%nonlinear kn_1 stiffness matrix
%....................................
Kn_1=zeros(sdof_q,sdof_q);

%....................................
%nonlinear kn_2 stiffness matrix
%....................................
Kn_2=zeros(sdof_q,sdof_q);

%....................................
%nonlinear kn_3 stiffness matrix
%....................................
Kn_3=zeros(sdof_q,sdof_q);

%....................................
%--------------------------------------------------------------------
% loop for computation and assembly of element matrices
%--------------------------------------------------------------------
% [point2,weight2]=feglqd2(nglx,ngly);              % call of
function for sampling points and weights
[point1,weight1]=feglqd2(nglx);
for iel=1:nel                 % loop for total number of elements
    for i=1:nnel
    nd(i)=nodes(iel,i); % extract connected node for (iel)-th
                        element
    xcoord(i)=gxcoord(nd(i),1);        % extract x value of node
  %  ycoord(i)=gycoord(nd(i),2);        % extract y value of node
    %eleng=ycoord(i)-xcoord(i);
```

```
     end
fd=zeros(edof_q,1);
 f=zeros(edof_q,1);
 fEm=zeros(edof_q,1);
  M=zeros(edof_q,1);
%  dff2d9=zeros(edof_q,1);
 dff2d2=zeros(edof_q,1);
k_q_q=zeros(edof_q,edof_q);

kf=zeros(edof_q,edof_q);
kf1=zeros(edof_q,edof_q);
kf2=zeros(edof_q,edof_q);
kf3=zeros(edof_q,edof_q);
kfnl3=zeros(edof_q,edof_q);
kfnl33=zeros(edof_q,edof_q);
kfnl=zeros(edof_q,edof_q);




M=zeros(edof_q,edof_q);
 dffEm=zeros(edof_q,1);%---diff load qm w r to Em

%disp=zeros(edof_q,1);disp1=zeros(edof_q,1);disp2=zeros(edof_q
,1);disp3=zeros(edof_q,1);disp4=zeros(edof_q,1);disp5=zeros(ed
of_q,1);disp6=zeros(edof_q,1);
% M=zeros(edof_q,edof_q);
%.........................................
%nonlinear kn_1 stiffness matrix
%.........................................
kn_1=zeros(edof_q,edof_q);

%.........................................
%nonlinear kn_2 stiffness matrix
%.........................................
kn_2=zeros(edof_q,edof_q);

%.........................................
%nonlinear kn_3 stiffness matrix
%.........................................
kn_3=zeros(edof_q,edof_q);

% k_g=zeros(edof_q,edof_q);
%.........................................
%--------------------------------------------------------------------------
% numarical integration
for intx=1:nglx
    x=point1(intx,1);                    % sampling point in x-axis
    wtx=weight1(intx,1);                   % weight in x-axis
%   for inty=1:ngly
%       y=point2(inty,2);                % sampling point in y-axis
%       wty=weight2(inty,2);               % weight in y-axis
```

```
      [shape,dhdr]=feisoq4beam4(nnel,x,h);            % call of
function to compute shape functions and
                         % derivatives of sampling points
jacob2=fejacob2beam4(nnel,dhdr,xcoord);   % call of function
to compute jacobian
detjacob=det(jacob2);                      % determination of
jacobian matrix
invjacob=inv(jacob2);              % inverse of jacobian matrix
[dhdx]=federiv2beam(nnel,dhdr,invjacob);  % call of function
to compute derivatives of
                                      % shape
functions w.r.t. physical coordinates

%%-------------------------------------------------------------
  [A,FW,Af,G]=a_gd(disp_w_e,ff_w_e,shape,dhdx,iel,s);
% compute shape function matrix (N)
%-------------------------------------------------------------
%%===========================================================
=====================
%    THERMAL  FORCES ,MOMENTS AND SHEAR FORCE
%===========================================================
=====================
c1=4/(3*h^2);
%[Bt,Bb1,Bs1,Bb2,Bs2,Bgg]=bmat_qt(nnel,shape,dhdx,h,s);
%[Bt,Bb1,Bs1,Bb2,Bs2,Bckg]=bckmat_q(nnel,shape,dhdx,h,i);
Bt1 =[-1 0 0 0 1 0 0 0];
%Bt1 =[-1 0 0 0 1 0 0 0];


% % ---- UDL RANDOM LOAD -------
% q=1000000.0;
% II=h^3/12;
%   q=qm*(Em*h^4)/(a^4);
%   I=(b*h^3)/12;
q=qm*(Em*I)/a^3;

% q=qm*(Em*II)/(a^3);
% qm=q*(a^4)/(Em*h^4);
% dqmdq=(a^4)/(Em*h^4)*10^(6);
% %dqdEm=qm*(1*h^4)/(a^4);
% dqd2=1*(Em*h^4)/(a^4);


% %dqmdEm=-q*(a^4)/(Em^2*h^4);
% dqd9=qm*(1*h^4)/(a^4);
%P=[0 q q 0]';
% dPdEm=[0 dqdEm 0 0]';
% dPd2=[0 dqd2 0 0]';
% dPd9=[0 0 dqd9 0 0 0 0]';


%-----------------
% % Reference force vector
```

```
% M=M+N'*m*N*wtx*wty*detjacob;
% f=Bg_q*q; %*wtx*detjacob;
% f=q*[0 leng/2 leng^2/12 0 0 leng/2 -leng^2/12 0 ]';

 f=(FW+q)*[0 leng/2 leng^2/12 0 0 leng/2 -leng^2/12 0 ]';
 fd=(FW+q)*[0 1 1 0 0 1 -1 0 ]';
% dff2d2=dff2d2+dqd2*[0 leng/2 leng^2/12 0 0 leng/2 -leng^2/12
0 ]';
%dff2d2=dqmdq*[0 leng/2 0 0 0 -leng/2 0 0 ]';
% fEm=fEm+ dqd9*[0 leng/2 leng^2/12 0 0 leng/2 -leng^2/12 0
]';

% FW
%%%%%%%%%%%%%%%%%%%%%%%%%%%%%%%%%%%%%%%%%%%%%%%%%%%%%%%%%%%%%%%%%%
%%%%%%
% STIFFNESS MATRIX
%%%%%%%%%%%%%%%%%%%%%%%%%%%%%%%%%%%%%%%%%%%%%%%%%%%%%%%%%%%%%%%%%%
%%%%%%

dN1dN1=1/s;dN2dN2=1/s;dN1dN2=-1/s;
N1N1=s/3;N1N2=s/6; N2N2=s/3;

dN1dN3=1/s; dN1dN5=-1/s; dN2dN3=-1/s; dN2dN5=1/s;
dN1dN4=0;dN1dN6=s/12;dN2dN4=0;dN2dN6=-s/12;
d2N3d2N3=12/s^3;d2N3d2N5=-12/s^3;d2N5d2N5=12/s^3;
N2dN3=-1/2;N1dN3=-1/2;N2dN5=1/2;N1dN5=1/2;
N1dN4=s/12; N2dN4=-s/12;N1dN6=-s/12;N2dN6=s/12;

dN3dN3=6/(5*s);dN3dN5=-6/(5*s);dN5dN5=6/(5*s);
dN4dN4=2*s/15;dN6dN6=2*s/15;dN4dN6=-s/30;
dN3dN4=1/10;dN3dN6=1/10;dN4dN5=-1/10;dN5dN6=-1/10;

dN1d2N4=1/s;dN1d2N6=-1/s;dN2d2N4=-1/s;dN2d2N6=1/s;
dN1d2N5=0; dN2d2N5=0; dN1d2N3=0; dN2d2N3=0;
N1d2N3=-1/s;N2d2N3=1/s;N2d2N5=-1/s;N1d2N5=1/s;

d2N4d2N4=4/s;d2N6d2N6=4/s;d2N4d2N6=2/s;
d2N3d2N4=6/s^2; d2N3d2N6=6/s^2;
d2N4d2N5=-6/s^2;d2N5d2N6=-6/s^2;

N1N3=7*s/20;N1N4=s^2/20;N1N5=3*s/20;N1N6=-s^2/30;
N2N3=13*s/20;N2N4=s^2/30;N2N5=7*s/20;N2N6=-s^2/20;

N3N3=13*s/35; N3N4=11*s^2/210; N3N5=9*s/70; N3N6=-13*s^2/420;
N4N4=s^3/105; N4N5=13*2/420; N4N6=-s^3/140;
N5N5=13*s/35; N5N6=-11*s^2/210;N6N6=s^3/105;
c1=4/(3*h^2);

% k_q_q  Matrix HSDT for Beam
% k11=A11*dN1dN1; k12=0; k13=-c1*E11*dN1d2N4;
k14=(B11-c1*E11)*dN1dN1;
```

```
% k15=A11*dN1dN2; k16=0; k17=-c1*E11*dN1d2N6;
k18=(B11-c1*E11)*dN1dN2;
% k21=0; k22=c1^2*H11*d2N3d2N3+(A22-
6*c1*D22+9*c1^2*F22)*dN3dN3; k23=c1^2*H11*d2N3d2N4+(A22-
6*c1*D22+9*c1^2*F22)*dN3dN4;
k24=(A22-6*c1*D22+9*c1^2*F22)*N1dN3;
% k25=0; k26=c1^2*H11*d2N3d2N5+(A22-
6*c1*D22+9*c1^2*F22)*dN3dN5; k27=c1^2*H11*d2N3d2N6+(A22-
6*c1*D22+9*c1^2*F22)*dN3dN6;
k28=(A22-6*c1*D22+9*c1^2*F22)*N2dN3;
% k31=k13; k32=k23; k33=(c1^2*H11)*d2N4d2N4+(A22-
6*c1*D22+9*c1^2*F22)*dN4dN4;
k34=(c1*(c1*H11-F11))*dN1d2N4+(A22-6*c1*D22+9*c1^2*F22)*N1dN4;
% k35=-c1*E11*dN2d2N4; k36=(c1^2*H11)*d2N4d2N5+(A22-
6*c1*D22+9*c1^2*F22)*dN4dN5; k37=(c1^2*H11)*d2N4d2N6+(A22-
6*c1*D22+9*c1^2*F22)*dN4dN6;
k38=(c1*(c1*H11-F11))*dN2d2N4+(A22-6*c1*D22+9*c1^2*F22)*N2dN4;
% k41=k14; k42=k24; k43=k34;
k44=(D11-2*c1*F11+c1^2*H11)*dN1dN1+(A22-
6*c1*D22+9*c1^2*F22)*N1N1;
% k45=(B11-c1*E11)*dN1dN2;k46=(c1*(c1*H11-F11))*dN1d2N5+(A22-
6*c1*D22+9*c1^2*F22)*N1dN5;
k47=(c1*(c1*H11-F11))*dN1d2N6+(A22-6*c1*D22+9*c1^2*F22)*N1dN6;
%
k48=(D11-2*c1*F11+c1^2*H11)*dN1dN2+(A22-
6*c1*D22+9*c1^2*F22)*N1N2;
% k51=k15; k52=k25; k53=k35;k54=k45; k55=A11*dN2dN2; k56=0;
k57=-c1*E11*dN2d2N6 ; k58=(B11-c1*E11)*dN2dN2;
% k61=k16; k62=k26; k63=k36; k64=k46; k65=k56;
k66=(c1^2*H11)*d2N5d2N5+(A22-6*c1*D22+9*c1^2*F22)*dN5dN5;
k67=(c1^2*H11)*d2N5d2N6+(A22-6*c1*D22+9*c1^2*F22)*dN5dN6;
k68=(c1*(c1*H11-F11))*dN2d2N5+(A22-6*c1*D22+9*c1^2*F22)*N2dN5;
% k71=k17; k72=k27; k73=k37; k74=k47; k75=k57; k76=k67;
k77=c1^2*H11*d2N6d2N6+(A22-6*c1*D22+9*c1^2*F22)*dN6dN6;
k78=(c1*(c1*H11-F11))*dN2d2N6+(A22-6*c1*D22+9*c1^2*F22)*N2dN6;
% k81=k18; k82=k28; k83=k38; k84=k48; k85=k58;
k86=k68;k87=k78;
k88=(D11-2*c1*F11+c1^2*H11)*dN2dN2+(A22-
6*c1*D22+9*c1^2*F22)*N2N2;


k11=A11*dN1dN1; k12=0; k13=-c1*E11*dN1d2N4;
k14=(B11-c1*E11)*dN1dN1;
k15=A11*dN1dN2; k16=0; k17=-c1*E11*dN1d2N6;
k18=(B11-c1*E11)*dN1dN2;
k21=0; k22=c1^2*H11*d2N3d2N3+(A22-6*c1*D22+9*c1^2*F22)*dN3dN3;
k23=(c1^2*H11*d2N3d2N4+(A22-6*c1*D22+9*c1^2*F22)*dN
3dN4)*10^(3); k24=((A22-6*c1*D22+9*c1^2*F22)*N1dN3)*10^(3);
k25=0; k26=(c1^2*H11*d2N3d2N5+(A22-
6*c1*D22+9*c1^2*F22)*dN3dN5);
k27=10^(3)*(c1^2*H11*d2N3d2N6+(A22-
6*c1*D22+9*c1^2*F22)*dN3dN6);
k28=10^(3)*((A22-6*c1*D22+9*c1^2*F22)*N2dN3);
```

```
k31=k13;  k32=k23;  k33=10^6*((c1^2*H11)*d2N4d2N4+(A22-
6*c1*D22+9*c1^2*F22)*dN4dN4);
k34=10^6*((c1*(c1*H11-F11))*dN1d2N4+(A22-
6*c1*D22+9*c1^2*F22)*N1dN4);
k35=-c1*E11*dN2d2N4;  k36=10^(3)*((c1^2*H11)*d2N4d2N5+(A22-
6*c1*D22+9*c1^2*F22)*dN4dN5);
k37=10^6*((c1^2*H11)*d2N4d2N6+(A22-
6*c1*D22+9*c1^2*F22)*dN4dN6);
k38=10^6*((c1*(c1*H11-F11))*dN2d2N4+(A22-
6*c1*D22+9*c1^2*F22)*N2dN4);
k41=k14;  k42=k24;  k43=k34;
k44=10^6*((D11-2*c1*F11+c1^2*H11)*dN1dN1+(A22-
6*c1*D22+9*c1^2*F22)*N1N1);
k45=(B11-c1*E11)*dN1dN2;k46=10^3*((c1*(c1*H11-
F11))*dN1d2N5+(A22-6*c1*D22+9*c1^2*F22)*N1dN5);
k47=10^6*(((c1*(c1*H11-F11))*dN1d2N6+(A22-
6*c1*D22+9*c1^2*F22)*N1dN6));
k48=10^6*((D11-2*c1*F11+c1^2*H11)*dN1dN2+(A22-
6*c1*D22+9*c1^2*F22)*N1N2);
k51=k15;  k52=k25;  k53=k35;k54=k45;  k55=A11*dN2dN2;  k56=0;
k57=-c1*E11*dN2d2N6 ;  k58=(B11-c1*E11)*dN2dN2;
k61=k16;  k62=k26;  k63=k36;  k64=k46;  k65=k56;
k66=((c1^2*H11)*d2N5d2N5+(A22-6*c1*D22+9*c1^2*F22)*dN5dN5);
k67=10^3*((c1^2*H11)*d2N5d2N6+(A22-
6*c1*D22+9*c1^2*F22)*dN5dN6);
k68=10^3*((c1*(c1*H11-F11))*dN2d2N5+(A22-
6*c1*D22+9*c1^2*F22)*N2dN5);
k71=k17;  k72=k27;  k73=k37;  k74=k47;  k75=k57;  k76=k67;
k77=10^6*(c1^2*H11*d2N6d2N6+(A22-6*c1*D22+9*c1^2*F22)*dN6dN6);
k78=10^6*((c1*(c1*H11-F11))*dN2d2N6+(A22-
6*c1*D22+9*c1^2*F22)*N2dN6);
k81=k18;  k82=k28;  k83=k38;  k84=k48;  k85=k58;  k86=k68;k87=k78;
k88=10^6*((D11-2*c1*F11+c1^2*H11)*dN2dN2+(A22-
6*c1*D22+9*c1^2*F22)*N2N2);


k_q_q=[k11 k12 k13 k14 k15 k16 k17 k18;
       k21 k22 k23 k24 k25 k26 k27 k28;
       k31 k32 k33 k34 k35 k36 k37 k38;
       k41 k42 k43 k44 k45 k46 k47 k48;
       k51 k52 k53 k54 k55 k56 k57 k58;
       k61 k62 k63 k64 k65 k66 k67 k68;
       k71 k72 k73 k74 k75 k76 k77 k78;
       k81 k82 k83 k84 k85 k86 k87 k88];



%%%%%%%%%%%%%%%%%%%%%%%%%%%%%%%%%%%%%%%%%%%%%%%%%%%%%%%%%%%%%%%
% Mass Matrix
%%%%%%%%%%%%%%%%%%%%%%%%%%%%%%%%%%%%%%%%%%%%%%%%%%%%%%%%%%%%%%%
%%%%%%%%%%%%%%%%%%%%%%%%%%%%%%%%%%%%%%%%%%%%%%%%%%%%%%%%%%%%%%%
% Mass Matrix
```

```
%%%%%%%%%%%%%%%%%%%%%%%%%%%%%%%%%%%%%%%%%%%%%%%%%%%%%%%%%%%%%%%%%%%%%%%
C1=1; C2= 4/(3*h^2);C4=C2;

N1N1=s/3;N1N2=s/6; N2N2=s/3;
N3N3=13*s/35; N3N4=11*s^2/210; N3N5=9*s/70; N3N6=-13*s^2/420;
N4N4=s^3/105; N4N5=13*s^2/420; N4N6=-s^3/140;
N5N5=13*s/35; N5N6=-11*s^2/210; N6N6=s^3/105;

dN1dN1=1/s;dN2dN2=1/s;dN1dN2=-1/s;

dN1dN4=0;dN1dN6=s/12;dN2dN4=0;dN2dN6=-s/12;
dN3dN3=6/(5*s);dN3dN5=-6/(5*s);dN5dN5=6/(5*s);
N2dN3=-1/2;N1dN3=-1/2;N2dN5=1/2;N1dN5=1/2;
N1dN4=s/12; N2dN4=-s/12;N1dN6=-s/12;N2dN6=s/12;


%%%%%%%%% (Winkler) linear  foundation stiffness matrix
%%%%%%
k11=N1N1; k12=N1N3; k13=N1N4; k14=N1N1; k15=N1N2; k16=N1N5;
k17=N1N6; k18=N1N2;
k21=N1N3; k22=N3N3; k23=N3N4; k24=N1N3; k25=N2N3; k26=N3N5;
k27=N3N6; k28=N2N3;
k31=k13; k32=k23; k33=N4N4; k34=N1N4; k35=N2N4; k36=N4N5;
k37=N4N6; k38=N2N4;
k41=k14; k42=k24; k43=k34; k44=N1N1; k45=N1N2;k46=N1N5;
k47=N1N6;k48=N1N2;
k51=k15; k52=k25; k53=k35;k54=k45; k55=N2N2; k56=N2N5;
k57=N2N6; k58=N2N2;
k61=k16; k62=k26; k63=k36; k64=k46; k65=k56; k66=N5N5;
k67=N5N6; k68=N2N5;
k71=k17; k72=k27; k73=k37; k74=k47; k75=k57;
k76=k67;k77=N6N6; k78=N2N6;
k81=k18; k82=k28; k83=k38; k84=k48; k85=k58; k86=k68;k87=k78;
k88=N2N2;

% k11=0; k12=0; k13=0; k14=0; k15=0; k16=0; k17=0; k18=0;
% k21=0; k22=N3N3; k23=N3N4; k24=0; k25=0; k26=N3N5;
k27=N3N6; k28=0;
% k31=k13; k32=k23; k33=N4N4; k34=0; k35=0; k36=N4N5;
k37=N4N6; k38=0;
% k41=k14; k42=k24; k43=k34; k44=0; k45=0; k46=0;
k47=0;k48=0;
% k51=k15; k52=k25; k53=k35;k54=k45; k55=0; k56=0; k57=0;
k58=0;
% k61=k16; k62=k26; k63=k36; k64=k46; k65=k56; k66=N5N5;
k67=N5N6; k68=0;
% k71=k17; k72=k27; k73=k37; k74=k47; k75=k57;
k76=k67;k77=N6N6; k78=0;
% k81=k18; k82=k28; k83=k38; k84=k48; k85=k58;
k86=k68;k87=k78; k88=0;
```

%

```
    kfl=K1*[k11 k12 k13 k14 k15 k16 k17 k18;
           k21 k22 k23 k24 k25 k26 k27 k28;
           k31 k32 k33 k34 k35 k36 k37 k38;
           k41 k42 k43 k44 k45 k46 k47 k48;
           k51 k52 k53 k54 k55 k56 k57 k58;
           k61 k62 k63 k64 k65 k66 k67 k68;
           k71 k72 k73 k74 k75 k76 k77 k78;
           k81 k82 k83 k84 k85 k86 k87 k88];

%%%%%%%%%% (Pasternak) shear  foundation stiffness matrix  %%%%%%
k11=dN1dN1;  k12=dN1dN3;  k13=dN1dN4;  k14=dN1dN1;  k15=dN1dN2;
k16=dN1dN5;  k17=dN1dN6;  k18=dN1dN2;
k21=dN1dN3;   k22=dN3dN3;  k23=dN3dN4;  k24=dN1dN3;k25=dN2dN3;
k26=dN3dN5;  k27=dN3dN6;  k28=dN2dN3;
k31=k13;  k32=k23;  k33=dN4dN4;  k34=dN1dN4;  k35=dN2dN4;
k36=dN4dN5;  k37=dN4dN6;  k38=dN2dN4;
k41=k14;  k42=k24;  k43=k34;  k44=dN1dN1;  k45=dN1dN2;
k46=dN1dN5;  k47=dN1dN6;  k48=dN1dN2;
k51=k15;  k52=k25;  k53=k35;k54=k45;  k55=dN2dN2;  k56=dN2dN5;
k57=dN2dN6;  k58=dN2dN2;
k61=k16;  k62=k26;  k63=k36;  k64=k46;  k65=k56;  k66=dN5dN5;
k67=dN5dN6;  k68=dN2dN5;
k71=k17;  k72=k27;  k73=k37;  k74=k47;  k75=k57;
k76=k67;k77=dN6dN6;  k78=dN2dN6;
k81=k18;  k82=k28;  k83=k38;  k84=k48;  k85=k58;  k86=k68;k87=k78;
k88=dN2dN2;

% k11=N1N1;  k12=N1dN3;  k13=N1dN4;  k14=N1N1;  k15=N1N2;
k16=N1dN5;  k17=N1dN6;  k18=N1N2;
% k21=N1dN3;   k22=dN3dN3;  k23=dN3dN4;  k24=N1dN3;k25=N2dN3;
k26=dN3dN5;  k27=dN3dN6;  k28=N2dN3;
% k31=k13;  k32=k23;  k33=dN4dN4;  k34=N1dN4;  k35=N2dN4;
k36=dN4dN5;  k37=dN4dN6;  k38=N2dN4;
% k41=k14;  k42=k24;  k43=k34;  k44=N1N1;  k45=N1N2;  k46=N1dN5;
k47=N1dN6;  k48=N1N2;
% k51=k15;  k52=k25;  k53=k35;k54=k45;  k55=N2N2;  k56=N2dN5;
k57=N2dN6;  k58=N2N2;
% k61=k16;  k62=k26;  k63=k36;  k64=k46;  k65=k56;  k66=dN5dN5;
k67=dN5dN6;  k68=N2dN5;
% k71=k17;  k72=k27;  k73=k37;  k74=k47;  k75=k57;
k76=k67;k77=dN6dN6;  k78=N2dN6;
% k81=k18;  k82=k28;  k83=k38;  k84=k48;  k85=k58;
k86=k68;k87=k78;  k88=N2N2;

% k11=0;  k12=0;  k13=0;  k14=0;  k15=0;  k16=0;  k17=0;  k18=0;
% k21=0;   k22=dN3dN3;  k23=dN3dN4;  k24=0;  k25=0;  k26=dN3dN5;
k27=dN3dN6;  k28=0;
% k31=k13;  k32=k23;  k33=dN4dN4;  k34=0;  k35=0;  k36=dN4dN5;
k37=dN4dN6;  k38=0;
```

```
% k41=k14; k42=k24; k43=k34; k44=0; k45=0; k46=0;
k47=0;k48=0;
% k51=k15; k52=k25; k53=k35;k54=k45; k55=0; k56=0; k57=0;
k58=0;
% k71=k17; k72=k27; k73=k37; k74=k47; k75=k57;
k76=k67;k77=dN6dN6; k78=0;
% k81=k18; k82=k28; k83=k38; k84=k48; k85=k58;
k86=k68;k87=k78; k88=0;

  kfs=K2*[k11 k12 k13 k14 k15 k16 k17 k18;
          k21 k22 k23 k24 k25 k26 k27 k28;
          k31 k32 k33 k34 k35 k36 k37 k38;
          k41 k42 k43 k44 k45 k46 k47 k48;
          k51 k52 k53 k54 k55 k56 k57 k58;
          k61 k62 k63 k64 k65 k66 k67 k68;
          k71 k72 k73 k74 k75 k76 k77 k78;
          k81 k82 k83 k84 k85 k86 k87 k88];

%%%%%%%%%%(winkler cubic nonlineraity) nonlinear foundation
stiffness matrix  %%%%%%
% k11=N1N1; k12=N1dN3; k13=N1dN4; k14=N1N1;
% k15=N1N2; k16=N1dN5; k17=N1dN6; k18=N1N2;
% k21=N1dN3; k22=dN3dN3+N3N3; k23=dN3dN4+N3N4; k24=N1dN3;
% k25=N2dN3; k26=dN3dN5+N3N5; k27=dN3dN6+N3N6; k28=N2dN3;
% k31=k13; k32=k23; k33=dN4dN4+N4N4; k34=N1dN4;
% k35=N2dN4; k36=dN4dN5+N4N5; k37=dN4dN6+N4N6; k38=N2dN4;
% k41=k14; k42=k24; k43=k34; k44=N1N1;
% k45=N1N2;k46=N1dN5; k47=N1dN6;k48=N1N2;
% k51=k15; k52=k25; k53=k35;k54=k45; k55=N2N2; k56=N2dN5;
k57=N2dN6; k58=N2N2;
% k61=k16; k62=k26; k63=k36; k64=k46; k65=k56;
k66=dN5dN5+N5N5; k67=dN5dN6+N5N6; k68=N2dN5;
% k71=k17; k72=k27; k73=k37; k74=k47; k75=k57;
k76=k67;k77=dN6dN6+N6N6; k78=N2dN6;
% k81=k18; k82=k28; k83=k38; k84=k48; k85=k58;
k86=k68;k87=k78; k88=N2N2;

k11=N1N1; k12=N1N3; k13=N1N4; k14=N1N1;k15=N1N2; k16=N1N5;
k17=N1N6; k18=N1N2;
k21=N1N3; k22=N3N3; k23=N3N4; k24=N1N3;k25=N2N3; k26=N3N5;
k27=N3N6; k28=N2N3;
k31=k13;  k32=k23; k33=N4N4; k34=N1N4; k35=N2N4; k36=N4N5;
k37=N4N6; k38=N2N4;
k41=k14;  k42=k24; k43=k34; k44=N1N1; k45=N1N2;k46=N1N5;
k47=N1N6; k48=N1N2;
k51=k15; k52=k25; k53=k35;k54=k45; k55=N2N2; k56=N2N5;
k57=N2N6; k58=N2N2;
k61=k16; k62=k26; k63=k36; k64=k46; k65=k56; k66=N5N5;
k67=N5N6; k68=N2N5;
k71=k17; k72=k27; k73=k37; k74=k47; k75=k57;
k76=k67;k77=N6N6; k78=N2N6;
```

```
k81=k18; k82=k28; k83=k38; k84=k48; k85=k58; k86=k68;k87=k78;
k88=N2N2;

% k11=0; k12=0; k13=0; k14=0; k15=0; k16=0; k17=0; k18=0;
% k21=0; k22=N3N3; k23=N3N4; k24=0; k25=0; k26=N3N5;
k27=N3N6; k28=0;
% k31=k13; k32=k23; k33=N4N4; k34=0; k35=0; k36=N4N5;
k37=N4N6; k38=0;
% k41=k14; k42=k24; k43=k34; k44=0; k45=0; k46=0;
k47=0;k48=0;
% k51=k15; k52=k25; k53=k35;k54=k45; k55=0; k56=0; k57=0;
k58=0;
% k61=k16; k62=k26; k63=k36; k64=k46; k65=k56; k66=N5N5;
k67=N5N6; k68=0;
% k71=k17; k72=k27; k73=k37; k74=k47; k75=k57;
k76=k67;k77=N6N6; k78=0;
% k81=k18; k82=k28; k83=k38; k84=k48; k85=k58;
k86=k68;k87=k78; k88=0;
%

    kfnl=Af^2*K3*[k11 k12 k13 k14 k15 k16 k17 k18;
        k21 k22 k23 k24 k25 k26 k27 k28;
        k31 k32 k33 k34 k35 k36 k37 k38;
        k41 k42 k43 k44 k45 k46 k47 k48;
        k51 k52 k53 k54 k55 k56 k57 k58;
        k61 k62 k63 k64 k65 k66 k67 k68;
        k71 k72 k73 k74 k75 k76 k77 k78;
        k81 k82 k83 k84 k85 k86 k87 k88];


%%%%%%%%%%%%%%%%%%%%%%%%%%%%%%%%%%%%%%%%%%%%%%%%%%%%%%%%%%%%%
% nonlinear stiffness martix
% %....................................

% kn_1= A*[0 A11*(6/(5*s)) 0 0 0 -A11*(6/(5*s)) 0 0;
%                0 0 0 0 0 0 0 0;
%                0 -c1*E11*(6/(5*s)) 0 0 0 -c1*E11*(-6/(5*s)) 0
0;
%                0 -(B11-c1*E11)*(-6/(5*s)) 0 0 0 -(B11-
c1*E11)*(6/(5*s)) 0 0;
%                0 -A11*(6/(5*s)) 0 0 0 A11*(6/(5*s)) 0 0 ;
%                0 0 0 0 0 0 0 0;
%                0 c1*E11*(6/(5*s)) 0 0 0 -c1*E11*(6/(5*s)) 0
0;
%                0 -(B11-c1*E11)*(6/(5*s)) 0 0 0 (B11-
c1*E11)*(6/(5*s)) 0 0];

%   kn_1= A*A11*[0 dN1dN3 dN1dN4 0 0 dN1dN5 dN1dN6 0;
%               dN1dN3 0 0 0 dN2dN3 0 0 0;
%               dN1dN4 0 0 0 dN2dN4 0 0 0;
%               0 0 0 0 0 0 0 0;
```

```
%                       0 dN2dN3 dN2dN4 0 0 dN2dN5 dN2dN6 0;
%                       dN1dN5 0 0 0 dN2dN5  0 0 0;
%                       dN1dN6 0 0 0 dN2dN6  0 0 0;
%                       0 0 0 0 0 0 0 0];

  kn_1= A*A11*[0 dN1dN3 dN1dN4 0 0 dN1dN5 dN1dN6 0;
               2*dN1dN3 0 0 0 2*dN2dN3  0 0 0;
               2*dN1dN4 0 0 0 2*dN2dN4  0 0 0;
               0 0 0 0 0 0 0 0;
               0 dN2dN3 dN2dN4 0 0 dN2dN5 dN2dN6 0;
               2*dN1dN5 0 0 0 2*dN2dN5  0 0 0;
               2*dN1dN6 0 0 0 2*dN2dN6  0 0 0;
               0 0 0 0 0 0 0 0];


    X=0; H11=0;
    kn_3= [0 0 0 0 0 0 0 0;
           0 (A11*(X+A^2)*dN3dN3-c1^2*H11*A*d2N3d2N3)
(A11*(X+A^2)*dN3dN4-c1^2*H11*A*d2N3d2N4) 0 0
(A11*(X+A^2)*dN3dN5-c1^2*H11*A*d2N3d2N5) (A11*A^2*dN3dN6-
c1^2*H11*A*d2N3d2N6) 0;
           0 (A11*(X+A^2)*dN3dN4-c1^2*H11*A*d2N3d2N4)
(A11*(X+A^2)*dN4dN4-c1^2*H11*A*d2N4d2N4) 0 0
(A11*(X+A^2)*dN4dN5-c1^2*H11*A*d2N4d2N5) (A11*A^2*dN4dN6-
c1^2*H11*A*d2N4d2N6) 0;
           0 0 0 0 0 0 0 0 ;
           0 0 0 0 0 0 0 0;
           0 (A11*(X+A^2)*dN3dN5-c1^2*H11*A*d2N3d2N5)
(A11*(X+A^2)*dN4dN5-c1^2*H11*A*d2N4d2N5) 0 0
(A11*(X+A^2)*dN5dN5-c1^2*H11*A*d2N5d2N5) (A11*A^2*dN5dN6-
c1^2*H11*A*d2N5d2N6) 0;
           0 (A11*(X+A^2)*dN3dN6-c1^2*H11*A*d2N3d2N6)
(A11*(X+A^2)*dN4dN6-c1^2*H11*A*d2N4d2N6) 0 0
(A11*(X+A^2)*dN5dN6-c1^2*H11*A*d2N5d2N6) (A11*A^2*dN6dN6-
c1^2*H11*A*d2N6d2N6) 0;
           0 0 0 0 0 0 0 0];

%

  %.............................................
  end
index_q=feeldof(nd,nnel,ndof_q);
index_qf=feeldof2f(nd,nnel,ndof_q);
index_q3f=feeldof3f(nd,nnel,ndof_q);
index_q4df2d2=feeldof4f(nd,nnel,ndof_q);
%-------------------------------------------------------------
% as^2semble element matrices
%-------------------------------------------------------------
K_q_q=feasmbl1(K_q_q,k_q_q,index_q);
Kf1=feasmbl1f(Kf1,kf1,index_qf);
Kfs=feasmbl5f(Kfs,kfs,index_qf);
```

```
Kfnl=feasmbl6f(Kfnl,kfnl,index_qf);
dFF2d2=feasmbl2q(dFF2d2,dff2d2,index_q);
FD=feasmbl2(FD,fd,index_q);
ff=feasmbl2(ff,f,index_q);
dffEm=feasmbl12Em(dffEm,fEm,index_q);


%.......................................
% non linear Kn_1 stiffness matrix
%.......................................
Kn_1=feasmbl1n1(Kn_1,kn_1,index_q);

% %...............................
% % non linear Kn_2 stiffness matrix
% %.................................
Kn_2=feasmbl1n2(Kn_2,kn_2,index_q);

% %...............................
% % non linear Kn_3 stiffness matrix
% %.................................
Kn_3=feasmbl1n3(Kn_3,kn_3,index_q);


%.......................................
%end

% [Ks1,Kg1]=feaplycs121(Kslm,Kg,bcdof);

 Ms=feasmbl1m(Ms,M,index_q);

end

%-----------------------------------------------------------------------
  %Ks=K_q_q-Kg+1/2*Kfnl+1/2*Kn_1+Kn_2+1/2*Kn_3;
%    Ks=K_q_q+Kf+1/2*Kfnl+1/2*Kn_1+Kn_2+1/2*Kn_3;
% Ks=K_q_q+1/2*Kn_1+Kn_2+1/2*Kn_3;
 %Ks=K_q_q+1/2*Kn_1*1.0e0+Kn_2*1.0e0+1/2*Kn_3*1.0e0;
  %Ks=K_q_q+(1/2)*Kn_1+(1/2)*Kn_2+(1/2)*Kn_3;
  Ks=K_q_q+(1/2)*Kn_1+(1/2)*Kn_3;
% Ks=K_q_q+Kfl+Kfs+0.25*Kfnl+(1/2)*Kn_1+(1/2)*Kn_3;
  % apply contraint and solve

%   [Kf]=feaplycs1(Kf,bcdof);
  [Ks1,ff]=feaplycs2(Ks,ff,bcdof);
  %[Ks1,F_t]=feaplycs33(Ks,F_t,bcdof);
 % [Ks1,F_t]=feaplycs2(Ks,F_t,bcdof);
%    [Kf1,ff]=feaplycs2f(Kf,ff,bcdof);
%    [Ks1,Ms1]=feaplycs2(Ks,Ms,bcdof);
%  disp=inv(Ks1)*(ff-F_t);
  disp=inv(Ks1)*(ff);
 % disp=inv(Ks1)*(F_t);
```

```
%  disp=-(0.5*Kn_1+ 0.5*Kn_3)*dispp;
  %-----------------------------------------------
  %-----------------------------------------------
  deflection_bar=(disp);    %*100*E2*h^3/(q*a^4);
% for ploting graph


j=1;
 for i=2:4:sdof_q
    disp_w(j)=disp(i,1);

    j=j+1;
end
en=max(disp_w);
if en==0
    disp_w=-disp_w;
    en=max(disp_w);
end
%%%%%%%%%%%%%%%%%%%  ff %%%%%%%%%%%%%%
j=1;
 for i=2:4:sdof_q
    ff_w(j)=FD(i,1);
    j=j+1;
end
enu=max(ff_w);
if en==0
    ff_w=-ff_w;
    enu=max(ff_w);
end
%.....Beam 2 No of element........
if nel==2

        disp_w_e(1,1)=deflection_bar(2);
        disp_w_e(1,2)=deflection_bar(6);

        disp_w_e(2,1)=deflection_bar(6);
        disp_w_e(2,2)=deflection_bar(10);

%.....Beam 4 No of element........
elseif nel==5

        disp_w_e(1,1)=deflection_bar(2);
        disp_w_e(1,2)=deflection_bar(6);

        disp_w_e(2,1)=deflection_bar(6);
        disp_w_e(2,2)=deflection_bar(10);

        disp_w_e(3,1)=deflection_bar(10);
        disp_w_e(3,2)=deflection_bar(14);

        disp_w_e(4,1)=deflection_bar(14);
```

```
        disp_w_e(4,2)=deflection_bar(18);

        disp_w_e(5,1)=deflection_bar(18);
        disp_w_e(5,2)=deflection_bar(22);


% %.....Beam 10 No of element.........
elseif nel==10

        disp_w_e(1,1)=deflection_bar(2);
        disp_w_e(1,2)=deflection_bar(6);

        disp_w_e(2,1)=deflection_bar(6);
        disp_w_e(2,2)=deflection_bar(10);
        disp_w_e(3,1)=deflection_bar(10);
        disp_w_e(3,2)=deflection_bar(14);

        disp_w_e(4,1)=deflection_bar(14);
        disp_w_e(4,2)=deflection_bar(18);
        disp_w_e(5,1)=deflection_bar(18);
        disp_w_e(5,2)=deflection_bar(22);

        disp_w_e(6,1)=deflection_bar(22);
        disp_w_e(6,2)=deflection_bar(26);

        disp_w_e(7,1)=deflection_bar(26);
        disp_w_e(7,2)=deflection_bar(30);

        disp_w_e(8,1)=deflection_bar(30);
        disp_w_e(8,2)=deflection_bar(34);

        disp_w_e(9,1)=deflection_bar(34);
        disp_w_e(9,2)=deflection_bar(38);

        disp_w_e(10,1)=deflection_bar(38);
        disp_w_e(10,2)=deflection_bar(42);

%.....Beam 16 No of element........
elseif nel==16
        disp_w_e(1,1)=deflection_bar(2);
        disp_w_e(1,2)=deflection_bar(6);

        disp_w_e(2,1)=deflection_bar(6);
        disp_w_e(2,2)=deflection_bar(10);

        disp_w_e(3,1)=deflection_bar(10);
        disp_w_e(3,2)=deflection_bar(14);

        disp_w_e(4,1)=deflection_bar(14);
        disp_w_e(4,2)=deflection_bar(18);
```

```
            disp_w_e(5,1)=deflection_bar(18);
            disp_w_e(5,2)=deflection_bar(22);

            disp_w_e(6,1)=deflection_bar(22);
            disp_w_e(6,2)=deflection_bar(26);

            disp_w_e(7,1)=deflection_bar(26);
            disp_w_e(7,2)=deflection_bar(30);

            disp_w_e(8,1)=deflection_bar(30);
            disp_w_e(8,2)=deflection_bar(34);

            disp_w_e(9,1)=deflection_bar(34);
            disp_w_e(9,2)=deflection_bar(38);

            disp_w_e(10,1)=deflection_bar(38);
            disp_w_e(10,2)=deflection_bar(42);

            disp_w_e(11,1)=deflection_bar(42);
            disp_w_e(11,2)=deflection_bar(46);

            disp_w_e(12,1)=deflection_bar(46);
            disp_w_e(12,2)=deflection_bar(50);

            disp_w_e(13,1)=deflection_bar(50);
            disp_w_e(13,2)=deflection_bar(54);

            disp_w_e(14,1)=deflection_bar(54);
            disp_w_e(14,2)=deflection_bar(58);

            disp_w_e(15,1)=deflection_bar(58);
            disp_w_e(15,2)=deflection_bar(62);

            disp_w_e(16,1)=deflection_bar(62);
            disp_w_e(16,2)=deflection_bar(66);

%.....Beam 20 No of element........
elseif nel==20
        disp_w_e(1,1)=deflection_bar(2);
        disp_w_e(1,2)=deflection_bar(6);

        disp_w_e(2,1)=deflection_bar(6);
        disp_w_e(2,2)=deflection_bar(10);

        disp_w_e(3,1)=deflection_bar(10);
        disp_w_e(3,2)=deflection_bar(14);

        disp_w_e(4,1)=deflection_bar(14);
        disp_w_e(4,2)=deflection_bar(18);

        disp_w_e(5,1)=deflection_bar(18);
        disp_w_e(5,2)=deflection_bar(22);
```

```
      disp_w_e(6,1)=deflection_bar(22);
      disp_w_e(6,2)=deflection_bar(26);

      disp_w_e(7,1)=deflection_bar(26);
      disp_w_e(7,2)=deflection_bar(30);

      disp_w_e(8,1)=deflection_bar(30);
      disp_w_e(8,2)=deflection_bar(34);

      disp_w_e(9,1)=deflection_bar(34);
      disp_w_e(9,2)=deflection_bar(38);

      disp_w_e(10,1)=deflection_bar(38);
      disp_w_e(10,2)=deflection_bar(42);

      disp_w_e(11,1)=deflection_bar(42);
      disp_w_e(11,2)=deflection_bar(46);

      disp_w_e(12,1)=deflection_bar(46);
      disp_w_e(12,2)=deflection_bar(50);

      disp_w_e(13,1)=deflection_bar(50);
      disp_w_e(13,2)=deflection_bar(54);

      disp_w_e(14,1)=deflection_bar(54);
      disp_w_e(14,2)=deflection_bar(58);

      disp_w_e(15,1)=deflection_bar(58);
      disp_w_e(15,2)=deflection_bar(62);

      disp_w_e(16,1)=deflection_bar(62);
      disp_w_e(16,2)=deflection_bar(66);

      disp_w_e(17,1)=deflection_bar(66);
      disp_w_e(17,2)=deflection_bar(70);

      disp_w_e(18,1)=deflection_bar(70);
      disp_w_e(18,2)=deflection_bar(74);

      disp_w_e(19,1)=deflection_bar(74);
      disp_w_e(19,2)=deflection_bar(78);

      disp_w_e(20,1)=deflection_bar(78);
      disp_w_e(20,2)=deflection_bar(82);
%.....Beam 24 No of element.........
elseif nel==24
      disp_w_e(1,1)=deflection_bar(2);
      disp_w_e(1,2)=deflection_bar(6);

      disp_w_e(2,1)=deflection_bar(6);
      disp_w_e(2,2)=deflection_bar(10);
```

```
disp_w_e(3,1)=deflection_bar(10);
disp_w_e(3,2)=deflection_bar(14);

disp_w_e(4,1)=deflection_bar(14);
disp_w_e(4,2)=deflection_bar(18);

disp_w_e(5,1)=deflection_bar(18);
disp_w_e(5,2)=deflection_bar(22);

disp_w_e(6,1)=deflection_bar(22);
disp_w_e(6,2)=deflection_bar(26);

disp_w_e(7,1)=deflection_bar(26);
disp_w_e(7,2)=deflection_bar(30);

disp_w_e(8,1)=deflection_bar(30);
disp_w_e(8,2)=deflection_bar(34);

disp_w_e(9,1)=deflection_bar(34);
disp_w_e(9,2)=deflection_bar(38);

disp_w_e(10,1)=deflection_bar(38);
disp_w_e(10,2)=deflection_bar(42);

disp_w_e(11,1)=deflection_bar(42);
disp_w_e(11,2)=deflection_bar(46);

disp_w_e(12,1)=deflection_bar(46);
disp_w_e(12,2)=deflection_bar(50);

disp_w_e(13,1)=deflection_bar(50);
disp_w_e(13,2)=deflection_bar(54);

disp_w_e(14,1)=deflection_bar(54);
disp_w_e(14,2)=deflection_bar(58);

disp_w_e(15,1)=deflection_bar(58);
disp_w_e(15,2)=deflection_bar(62);

disp_w_e(16,1)=deflection_bar(62);
disp_w_e(16,2)=deflection_bar(66);

disp_w_e(17,1)=deflection_bar(66);
disp_w_e(17,2)=deflection_bar(70);

disp_w_e(18,1)=deflection_bar(70);
disp_w_e(18,2)=deflection_bar(74);

disp_w_e(19,1)=deflection_bar(74);
disp_w_e(19,2)=deflection_bar(78);
```

```
        disp_w_e(20,1)=deflection_bar(78);
        disp_w_e(20,2)=deflection_bar(82);

        disp_w_e(21,1)=deflection_bar(82);
        disp_w_e(21,2)=deflection_bar(86);

        disp_w_e(22,1)=deflection_bar(86);
        disp_w_e(22,2)=deflection_bar(90);

        disp_w_e(23,1)=deflection_bar(90);
        disp_w_e(23,2)=deflection_bar(94);

        disp_w_e(24,1)=deflection_bar(94);
        disp_w_e(24,2)=deflection_bar(98);

        %.....Beam 30 No of element........
elseif nel==30
    ff=FD;
        disp_w_e(1,1)=deflection_bar(2);
        disp_w_e(1,2)=deflection_bar(6);

        ff_w_e(1,1)=ff(2);
        ff_w_e(1,2)=ff(6);

        disp_w_e(2,1)=deflection_bar(6);
        disp_w_e(2,2)=deflection_bar(10);

        ff_w_e(2,1)=ff(6);
        ff_w_e(2,2)=ff(10);

        disp_w_e(3,1)=deflection_bar(10);
        disp_w_e(3,2)=deflection_bar(14);

        ff_w_e(3,1)=ff(10);
        ff_w_e(3,2)=ff(14);

        disp_w_e(4,2)=deflection_bar(18);
                disp_w_e(4,1)=deflection_bar(14);

        ff_w_e(4,1)=ff(14);
        ff_w_e(4,2)=ff(18);

        disp_w_e(5,1)=deflection_bar(18);
        disp_w_e(5,2)=deflection_bar(22);

        ff_w_e(5,1)=ff(18);
        ff_w_e(5,2)=ff(22);

        disp_w_e(6,1)=deflection_bar(22);
        disp_w_e(6,2)=deflection_bar(26);
```

```
ff_w_e(6,1)=ff(22);
ff_w_e(6,2)=ff(26);

disp_w_e(7,1)=deflection_bar(26);
disp_w_e(7,2)=deflection_bar(30);

ff_w_e(7,1)=ff(26);
ff_w_e(7,2)=ff(30);

disp_w_e(8,1)=deflection_bar(30);
disp_w_e(8,2)=deflection_bar(34);

ff_w_e(8,1)=ff(30);
ff_w_e(8,2)=ff(34);

disp_w_e(9,1)=deflection_bar(34);
disp_w_e(9,2)=deflection_bar(38);


ff_w_e(9,1)=ff(34);
ff_w_e(9,2)=ff(38);

disp_w_e(10,1)=deflection_bar(38);
disp_w_e(10,2)=deflection_bar(42);

ff_w_e(10,1)=ff(38);
ff_w_e(10,2)=ff(42);

disp_w_e(11,1)=deflection_bar(42);
disp_w_e(11,2)=deflection_bar(46);

ff_w_e(11,1)=ff(42);
ff_w_e(11,2)=ff(46);


disp_w_e(12,1)=deflection_bar(46);
disp_w_e(12,2)=deflection_bar(50);

ff_w_e(12,1)=ff(46);
ff_w_e(12,2)=ff(50);

disp_w_e(13,1)=deflection_bar(50);
disp_w_e(13,2)=deflection_bar(54);

ff_w_e(13,1)=ff(50);
ff_w_e(13,2)=ff(54);

disp_w_e(14,1)=deflection_bar(54);
disp_w_e(14,2)=deflection_bar(58);

ff_w_e(14,1)=ff(54);
```

```
        ff_w_e(14,2)=ff(58);

        disp_w_e(15,1)=deflection_bar(58);
        disp_w_e(15,2)=deflection_bar(62);

        ff_w_e(15,1)=ff(58);
        ff_w_e(15,2)=ff(62);

        disp_w_e(16,1)=deflection_bar(62);
        disp_w_e(16,2)=deflection_bar(66);

        ff_w_e(16,1)=ff(62);
        ff_w_e(16,2)=ff(66);

        disp_w_e(17,1)=deflection_bar(66);
        disp_w_e(17,2)=deflection_bar(70);

        ff_w_e(17,1)=ff(66);
        ff_w_e(17,2)=ff(70);

        disp_w_e(18,1)=deflection_bar(70);
        disp_w_e(18,2)=deflection_bar(74);

        ff_w_e(18,1)=ff(70);
        ff_w_e(18,2)=ff(74);

        disp_w_e(19,1)=deflection_bar(74);
        disp_w_e(19,2)=deflection_bar(78);

         ff_w_e(19,1)=ff(74);
        ff_w_e(19,2)=ff(78);

        disp_w_e(20,1)=deflection_bar(78);
        disp_w_e(20,2)=deflection_bar(82);

        ff_w_e(20,1)=ff(78);
        ff_w_e(20,2)=ff(82);

        disp_w_e(21,1)=deflection_bar(82);
        disp_w_e(21,2)=deflection_bar(86);

        ff_w_e(21,1)=ff(82);
        ff_w_e(21,2)=ff(86);

        disp_w_e(22,1)=deflection_bar(86);
        disp_w_e(22,2)=deflection_bar(90);


        ff_w_e(22,1)=ff(86);
        ff_w_e(22,2)=ff(90);
```

```
disp_w_e(23,1)=deflection_bar(90);
disp_w_e(23,2)=deflection_bar(94);

ff_w_e(23,1)=ff(90);
ff_w_e(23,2)=ff(94);

disp_w_e(24,1)=deflection_bar(94);
disp_w_e(24,2)=deflection_bar(98);

ff_w_e(24,1)=ff(94);
ff_w_e(24,2)=ff(98);

disp_w_e(25,1)=deflection_bar(98);
disp_w_e(25,2)=deflection_bar(102);


ff_w_e(25,1)=ff(98);
ff_w_e(25,2)=ff(102);

disp_w_e(26,1)=deflection_bar(102);
disp_w_e(26,2)=deflection_bar(106);

ff_w_e(26,1)=ff(102);
ff_w_e(26,2)=ff(106);

disp_w_e(27,1)=deflection_bar(106);
disp_w_e(27,2)=deflection_bar(110);

ff_w_e(27,1)=ff(106);
ff_w_e(27,2)=ff(110);

disp_w_e(28,1)=deflection_bar(110);
disp_w_e(28,2)=deflection_bar(114);

ff_w_e(28,1)=ff(110);
ff_w_e(28,2)=ff(114);

disp_w_e(29,1)=deflection_bar(114);
disp_w_e(29,2)=deflection_bar(118);

ff_w_e(29,1)=ff(114);
ff_w_e(29,2)=ff(118);

disp_w_e(30,1)=deflection_bar(118);
disp_w_e(30,2)=deflection_bar(122);

ff_w_e(30,1)=ff(118);
ff_w_e(30,2)=ff(122);

end
```

```
 for i=1:nel
     for j=1:2
         no=nodes(i,j);
%            disp_w_e(i,j)=(disp_w_e(i,j)*Wmax*h)/en;
         ff_w_e(i,j)=(ff_w_e(i,j)*q)/enu;

%     ff_w_e(i,j)=enu/(ff_w_e(i,j)*q);

     end
 end
 if nel==30
     delect(sh)=deflection_bar(62)/h %*100*h^3*Em/(q*a^4)
%.....30 element beam
     qq(sh)=ff(66)
    %delect(sh)=deflection_bar(62)*100*h/(alpham*(Tc-Tm)*a^2)
%.....30 element beam
 elseif nel==24
     delect(sh)=deflection_bar(50)*100*h^3*Em/(q*a^4) %.....24
element beam
 elseif nel==20
     delect(sh)=deflection_bar(42)*100*h^3*Em/(q*a^4) %.....24
element beam
 elseif nel==16
     delect(sh)=deflection_bar(34)*100*h^3*Em/(q*a^4) %.....16
element beam
     %delect(sh)=deflection_bar(66)/h;%*100*h^3*Em/(q*a^4)
%.....16 element beam
 elseif nel==10
     delect(sh)=deflection_bar(22)*100*h^3*Em/(q*a^4) %.....11
element beam
 elseif nel==5
     delect(sh)=deflection_bar(10)*100*h^3*Em/(q*a^4) %.....5
element beam
 elseif nel==2
     delect(sh)=deflection_bar(6)*100*h^3*Em/(q*a^4) %.....2
element beam
 end
%     delect(sh)=deflection_bar(98)
    if (sh~=1 & abs(delect(sh)-delect(sh-1))<=(1e-2))

        delection_nonlinear1(pp)=(delect(sh))   %*100*E2*h^3/
(q*a^4)

  break
  end
 end
 end
 end
end

  if zz==1
     delection_linear=delection_nonlinear1;
```

```
  elseif zz==2
     delection_nonlinear=delection_nonlinear1;
  end
end
%for Vcn=0.12
% delection_linear=[0 1.56781 2.40469 3.03235 3.58156 4.05231
4.47075 4.86304];
%for Vcn=0.17
delection_linear=[0 1.39 1.99 2.37 2.6 2.9 3.11];


plot(value_q,delection_linear,'b',value_q,delection_
nonlinear,'r');
  legend('Shen & Xiang (2013)','Present')
```

REFERENCES

Clough, Ray. 1960. "Original formulation of the finite element method, Finite Elements in Analysis and Design." 7 (2): 89–101. https://doi.org/10.1016/0168-874X(90)90001-U.

Courant, Richard. 1943. "Variational methods for the solution of problems of equilibrium and vibrations." Bulletin of the American Mathematical Society 49: 1–23. Doi: 10.1090/S0002-9904-1943-07818-4

Hrennikoff, Alexander. 1941. "Solution of problems of elasticity by the framework method." Journal of Applied Mechanics 8(4): A169–A175. https://doi.org/10.1115/1.4009129

Lal, Achchhe and Kanif Markad. 2018. "Deflection and stress behaviour of multi-walled carbon nanotube reinforced laminated composite beams." Computers and Concrete 22 (6): 501–514. https://doi.org/10.12989/cac.2018.22.6.501

Lal, Achchhe and Kanif Markad. 2021. "Thermal post buckling analysis of smart SMA hybrid sandwich composite plate." Polymers and Polymer Composites 29(9): S344–S360. doi:10.1177/09673911211001276.

Markad, Kanif, and Achchhe Lal. 2021. "Experimental investigation of shape memory polymer hybrid nanocomposites modified by carbon fiber reinforced multi-walled carbon (MWCNT)." Materials Research Express 8: 105015. Doi: 10.1088/2053-1591/ac2fcc.

Markad, K., Das Vivek, and A. Lal. 2022. "Deflection and stress analysis of piezoelectric laminated composite plate under variable polynomial transverse loading." AIP Advances 12 (8): 085024. https://doi.org/10.1063/5.0104568

Tiwari, Nilesh, and AbdulHafiz A. Shaikh. 2019. "Flexural analysis of thermally actuated fiber reinforced shape memory polymer composite." Advances in Materials Research, 8(4): 337–359. https://doi.org/10.12989/amr.2019.8.4.337

Tiwari, Nilesh, and AbdulHafiz A. Shaikh. 2021. "Micro Buckling of Carbon Fiber in Triple Shape Memory Polymer Composites under Bending in Glass Transition Regions." Materials Today: Proceedings 44: 4744–8.

Wilson, Taylor, Doherty, and Ghaboussi. 1973. Incompatible displacement models. In Numerical Computational Methods in Structural Mechanics, edited by S. J. Fenves et al. New York: Academic Press.

Appendix 1
Case Study: Modeling of Shape Memory Polymer

In Eq. A1.1, a mathematical model was suggested to determine the storage modulus $E(T)$ of polymers across a broad temperature range from room temperature to T_g temperature while keeping a steady frequency. Similarly, Eq. A1.2 gives Poisson's ratio.

$$E(T) = (E_1 - E_2)e^{\left(-\left(\frac{T}{T_\beta}\right)^{m1}\right)} + (E_2 - E_3)e^{\left(-\left(\frac{T}{T_g}\right)^{m2}\right)} + E_3 e^{\left(-\left(\frac{T}{T_f}\right)^{m3}\right)} \qquad \text{(A1.1)}$$

$$\upsilon_m = \upsilon_g \lambda_g + \upsilon_r (1 - \lambda_g) \text{ and } \lambda_g = 1 - \frac{1}{1 + e^{[-(T-T_m)/Z]}} \qquad \text{(A1.2)}$$

The present study includes three temperature zones, namely β, glass, and flow region transition temperatures, which are shown by T_β, T_g, T_f, respectively, and the elastic modulus corresponds to these temperatures are E_1, E_2, E_3. Weibull coefficient expressed by the letter m. Eq. A1.1 can be simplified into Eq. A1.3. This equation considers the value of T_g, that is greater than T_β but lower than T_f.

$$E(T) = (E_2 - E_3).exp\left(-\left(\frac{T}{T_g}\right)^{m2}\right) + E_3 \qquad \text{(A1.3)}$$

Figure A.1 shows the behavior of the matrix storage modulus over the glass transition zone as a function of dynamic temperature fluctuations. Since the loss modulus is small in comparison to the storage modulus, the storage modulus has been used as Young's modulus in the ensuing analyses (Figure A.2).

In the present analysis, T_g is taken as 341 K and remains fixed unless stated. But it is also important to know the effect of variations in T_g for different matrix polymers, which may vary from 305 to 345 K. Figure A.3 shows the effect of dynamic temperature variation over E_m for the above-mentioned T_g. If you heat up crystals, the connections between them weaken. Since intermolecular interactions are feeble, the same tension will result in a much greater stretch, and crystals will not return to their original location. Therefore, elevated temperatures cause a loss of flexibility. With the increase in temperature, the modulus of the polymer significantly decreases, and simultaneous changes are observed in shear modulus and Poisson ratio as well. At room temperature, the material is stiff, and during the programming process, it is in a soft state. At high temperatures, the liquid crystals become disordered, and the material relaxes. But when cooled, the liquid crystals take on a specific orientation programmed into

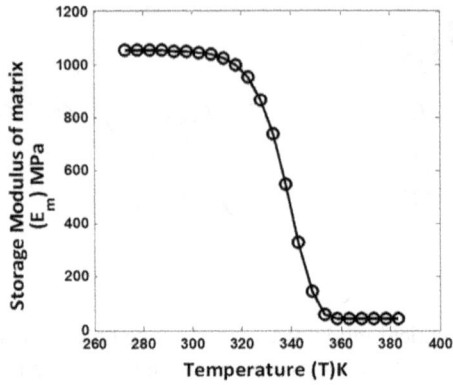

FIGURE A.1   Storage modulus of the shape memory polymer.

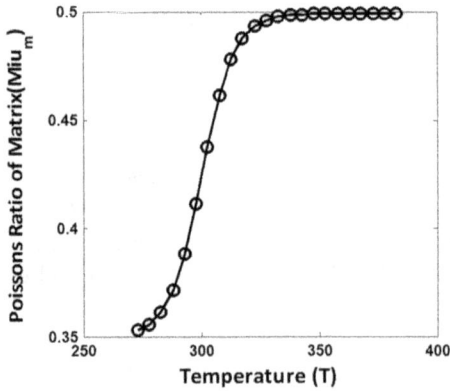

FIGURE A.2   Poisson's ratio of the shape memory polymer.

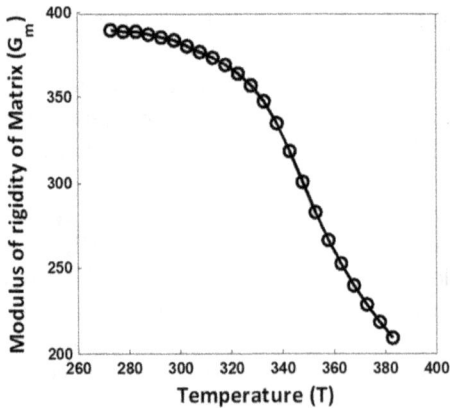

FIGURE A.3   Modulus of rigidity of the shape memory polymer.

the material by the researchers. This orientation holds the material in a specific shape. The low-temperature phase is called Martensite. But if we increase the temperature, the atomic alignment changes into another phase called Austenite. It is the high-temperature phase. In this phase, only one can set specific required shape to the material. ($T_g = 68 + 273$; $T_m = 27.5 + 273$; $E_{m2} = 1{,}053.41$; $E_{m3} = 44.11$; $m_2 = 41$; $z_m = 7$)

MATLAB CODE: A1 TEMPERATURE-DEPENDENT POLYMER MATRIX PROPERTY

```
clear all
close all
clc
value_T=[273:5:383];
for kk=1:23
    T=value_T(1,kk);
Tg=68+273;
Tm=27.5+273;
Em2=1053.41;
Em3=44.11;
m2=41;
zm=7;
Em(kk)=[(Em2-Em3)*exp(-(T/Tg)^(m2))]+Em3;
Miu_g=0.35;
Miu_r=0.499;
Fg(kk)=1-(1+exp(-(T-Tm)/zm))^(-1);
Miu_m(kk)=(Miu_g*Fg(kk))+(Miu_r*(1-Fg(kk)));
Gm(kk)=Em/(2*(1+Miu_m));
end

%% Plotting graph:
% hold on
% plot(value_T,Em,'b');
% hold on
% ylabel('Storage Modulus of matrix(E_m) MPa')
% xlabel('Temperature (T)K')
% grid on;

% hold on
% plot(value_T,Miu_m,'b');
% hold on
% ylabel('Poissons Ratio of Matrix(Miu_m)')
% xlabel('Temperature (T)')
% grid on;

% hold on
% plot(value_T,Gm,'b');
% hold on
% ylabel('Modulus of rigidity of Matrix (G_m)')
% xlabel('Temperature (T)')
% grid on;
```

Appendix 2
Case Study: Modeling of Shape Memory Polymer Composites

There are various methods of evaluating the material properties of two-phase and multiphase composites that involve the Voigt model, the Halpin–Tsai model, the Mori–Tanaka model, and many more. Here we are particularly focused on the evaluation of the effective material properties of the considered composites by the Halpin–Tsai approach, along with the analytical formulation of the Halpin–Tsai approach and the Mori–Tanaka approach.

Figure A2.1 shows the effective elastic properties of shape memory polymer composites. During the analysis, the fiber volume fraction is taken at 70% and the matrix at 30%. With the increment in temperature, the material first shows a glassy state, then undergoes a glass transition region, and then goes into a rubbery and flow region, respectively, based on the type of polymer. The material properties of the matrix are estimated as described in Appendix A2.

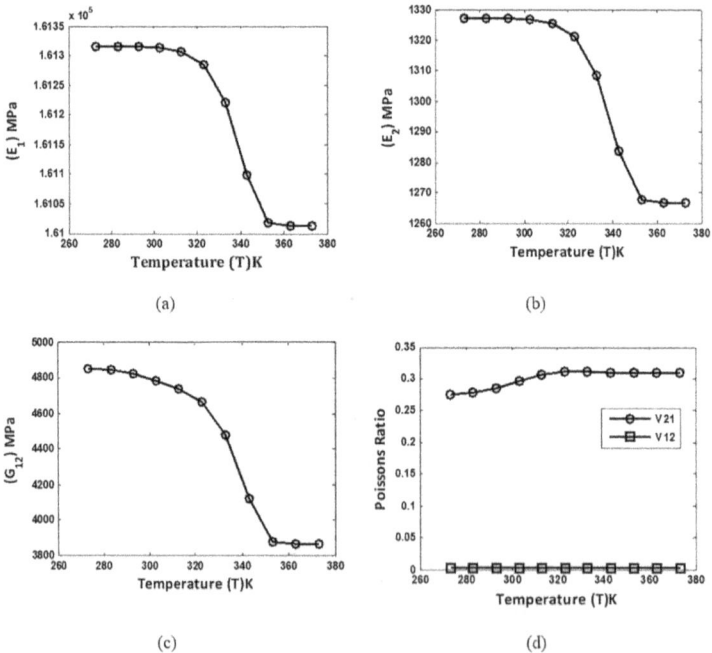

(a)

(b)

(c)

(d)

FIGURE A2.1 Effect of temperature variation over (a) Longitudinal modulus E_1, (b) Transverse modulus E_2, (c) Shear modulus G_{12}, and (d) Poisson's ratio of SMPC.

MATLAB CODE: A.2 TEMPERATURE-DEPENDENT PROPERTY
OF SHAPE MEMORY POLYMER COMPOSITE

```
clc
clear all
close all
format long

value_T=[273:10:373];
for kk=1:11
    T=value_T(1,kk);

Tg=68+273;
Tm=27.5+273;
Em2=1053.41;
Em3=44.11;
m2=41;
zm=7;

Em(kk)=[(Em2-Em3)*exp(-(T/Tg)^(m2))]+Em3;
Miu_g=0.35;
Miu_r=0.499;
Fg=1-(1+exp(-(T-Tm)/zm))^(-1);
Miu_m=(Miu_g*Fg)+(Miu_r*(1-Fg));
Gm=Em/(2*(1+Miu_m));

Miu_g=0.35;
Miu_r=0.499;
Fg=1-(1+exp(-(T-Tm)/zm))^(-1);
Miu_m=(Miu_g*Fg)+(Miu_r*(1-Fg));
Gm=Em(kk)/(2*(1+Miu_m));

%% Fiber property
Ef1=230e3;
Ef2=8.2e3;
Gf12=27.3e3;
Miu_f=0.25;
aplha_f1=-8.3e-7;
aplha_f2=10e-6;
aplha_f12=0;

%% Composite property
Vf=0.70; %input ('Fiber volume fraction=');
Vm=1-Vf;
EE_1_2=(Ef2*Em(kk))*((Ef2*Vm)+(Em(kk)*Vf))^(-1);
EE_2_2=(Ef2*Vf)+(Em*Vm);
Miu_1_21=(Miu_f*Vf)+(Miu_m*Vm);
Miu_2_21=[(Miu_f*Vf*Ef2)+(Miu_m*Vm*Em(kk))]*[(Ef2*Vf)+(Em(kk)
*Vm)]^(-1);
G_1_12(kk)=(Gf12*Gm)*[(Gf12*Vm)+(Gm*Vf)]^(-1);
G_2_12(kk)=(Gf12*Vf)+(Gm*Vm);
```

```
cc=0.2;

E1=(Ef1*Vf)+(Em*Vm);
E2=((1-cc)*EE_1_2)+(cc*EE_2_2);
G12=((1-cc)*G_1_12)+(cc*G_2_12);
G13=G12;
G23=G12;
V21(kk)=((1-cc)*Miu_1_21)+(cc*Miu_2_21);
V12=V21*(E2(kk)/E1(kk));
Miu_21=(Miu_m*Vm)+(Miu_f*Vf);
end
if T<Tg
    alpham=1.25*1.0e-4;
else T>Tg
    alpham=2.52*1.0e-4;
end
alp1=(alpham*Em*Vm+lpha_f1*Ef1*Vf)/(Em*Vm+Ef1*Vf);
alp2=(alpham*(1+Miu_m)*Vm)+(lpha_f2*(1+Miu_f)*Vf)-(alp1*Miu_21);
alp12=0;

%% Plotting Figures:
% figure(1)
% plot(value_T,E1,'b');
% ylabel('Storage Modulus (E_1) Mpa')
% xlabel('Temperature (T)K')
% grid on;

% figure(2)
% plot(value_T,E2,'b');
% ylabel('Storage Modulus of SMPC(E_2) Mpa')
% xlabel('Temperature (T)K')
% grid on;

% figure(3)
% plot(value_T,G12,'b');
% ylabel('Shear Modulus of SMPC(G_1_2) Mpa')
% xlabel('Temperature (T)K')
% grid on;

% figure(4)
% plot(value_T,V21,'b',value_T,V12,'r' );
% ylabel('Poissons Ratio')
% xlabel('Temperature (T)K')
% legend('V21','V12')
% grid on;

% figure(5)
% plot(value_T,V12,'b' );
% ylabel('Poissons Ratio V12')
% xlabel('Temperature (T)K')
% grid on;
```

Appendix 3
Case Study: Mechanical Properties of the Shape Memory Hybrid Composite Using the Halpin–Tsai Model

The modified matrix is a combination of matrix and single-walled carbon nanotubes. This modified matrix is further utilized with carbon fiber to form a three-phase composite material. Eqs. A3.1–A3.7 express the evaluation of the effective material properties of the three-phase composites. Fiber material parameters are taken as, $E_{f1} = 2{,}300\,\text{GPa}$, $E_{f2} = 8.2\,\text{GPa}$, $G_{f12} = 27.3\,\text{GPa}$, $\mu_f = 0.25$, and $C = 0.2$.

$$
E_c = \left\{ \left(\frac{3}{8}\right) \left[\frac{+2\left(\dfrac{L}{t_{NT}}\right)\dfrac{\left(\dfrac{N_w t_{NT} E_{NT}}{(N_w-1)h_{in}+t_{NT}}/E_{MT}\right)-1}{\left(\dfrac{N_w t_{NT} E_{NT}}{(N_w-1)h_{in}+t_{NT}}/E_{MT}\right)+2(L/t)}}{1-\dfrac{\left(\dfrac{N_w t_{NT} E_{NT}}{(N_w-1)h_{in}+t_{NT}}/E_{MT}\right)-1}{\left(\dfrac{N_w t_{NT} E_{NT}}{(N_w-1)h_{in}+t_{NT}}/E_{MT}\right)+2(L/t)} V_{cnt}} V_{cnt} \right] E_m \right.
$$

$$
\left. + \left(\frac{5}{8}\right) \left[\frac{1+2\dfrac{\left(\dfrac{N_w t_{NT} E_{NT}}{(N_w-1)h_{in}+t_{NT}}/E_{MT}\right)-1}{\left(\dfrac{N_w t_{NT} E_{NT}}{(N_w-1)h_{in}+t_{NT}}/E_{MT}\right)+2(1)} V_{cnt}}{1-\dfrac{\left(\dfrac{N_w t_{NT} E_{NT}}{(N_w-1)h_{in}+t_{NT}}/E_{MT}\right)-1}{\left(\dfrac{N_w t_{NT} E_{NT}}{(N_w-1)h_{in}+t_{NT}}/E_{MT}\right)+2(1)} V_{cnt}} \right] E_m \right\} \tag{A3.1}
$$

$$
E^1 = \frac{E_{f2}E_c}{E_{f2}v_m + E_c v_f}, \qquad E_{c1} = E_{f1}v_f + E_c v_m \tag{A3.2}
$$

$$E_{c2}^2 = E_{f2}v_f + E_c v_m, \quad E_{c2} = (1-C)E_{c2}^1 + CE_{c2}^2 \qquad (A3.3)$$

$$\mu_{c21}^1 = \mu_f v_f + \mu_m v_m, \quad \mu_{c21} = (1-C)\mu_{c21}^1 + C\mu_{c21}^2 \qquad (A3.4)$$

$$\mu_{c21}^2 = \frac{\mu_f E_{f2}v_f + \mu_m E_c v_m}{E_{f2}v_f + E_c v_m}, \quad \mu_{c12} = \mu_{c21}\frac{E_{c2}}{E_{c1}} \qquad (A3.5)$$

$$G_{c12}^1 = \frac{G_{f12}G_m}{G_{f12}v_m + G_m v_f} \qquad (A3.6)$$

$$G_{c12}^2 = G_{f12}v_f + G_m v_m \text{ and } G_{c12} = (1-C)G_{c12}^1 + CG_{c12}^2 \qquad (A3.7)$$

E_c = modified matrix. The 1 and 2 suffixes show properties of matrix and fiber in the longitudinal and transverse directions, respectively.

MATLAB CODE: A3 SMPHC MATERIAL PROPERTY EVALUATION BY HALPIN–TSAI MODEL

```
%    Haplin Tsai-Model %
%    Three Phase = (matrix+CNT)+ Fiber       %


clear
clc

T=input ('input remeperature=')
T0=300;    %(reference temepraturei.e. room temperature=27oC)
deltaT=T-T0;
Em=2.5E9;  % Temperature independent matrix
% T=300;       %input ('input temperature=')
% T0=300;    %(reference temperature)
% deltaT=T-T0;
% % Em=2.5E9;
% Em=(3.52-0.0034*deltaT)*1.0E9; % Temperature dependent marix
numm=0.3;rhom=1180; % matrix
%% SWCNT
Nw=1;       %input ('number of CNTs=')
Ecnt=1000E9;nucnt=0.28;rhocnt=1300; % SWCNT fiber
Lcnt=2.0E-7; %  length of SWCNT
dcnt = 1.0E-9; % diamater of SWCNT
tcnt=0.335e-9; % thickness of SWCNT
hin=1.5*tcnt;
n=10;m=10;
acc=0.141E-9;
Dcnt=sqrt(3*(n^2+m^2+n*m))*pi^(-1)*acc;
Dso=Dcnt+tcnt;
Dsi=Dcnt-tcnt;
Dmo=Dso+2*(Nw-1)*hin;
```

```
Dmi=Dsi;
%%%%%%%%%%% multiwall CNT %%%%%%%%%%%%%%
Emw=(Nw*tcnt*Ecnt)*((Nw-1)*hin+tcnt)^(-1);% for multiwall
%%%%%%%%%%%%%%%%%%%%%%%%%%%%%%%%
 Vcnt=input ('volume fraction of CNTs=')
 etaL=2*(Lcnt*(Dmo-Dmi)^(-1));
 etaT=2;
 AAn=(Emw*(Em)^(-1)-1);
 AAd=Emw*(Em)^(-1)+etaL;
 nueL=AAn/AAd;
 BBn=Emw*(Em)^(-1)-1;
 BBd=Emw*(Em)^(-1)+etaT;
 nueT=BBn/BBd;
EL=((1+nueL*etaL*Vcnt)/(1-nueL*Vcnt))*Em;
ET=(1+nueL*etaT*Vcnt)/(1-nueT*Vcnt)*Em;

EC=[(3/8)*EL+(5/8)*ET];

xi=[2 1];
Ef=69E9;nuf=0.22;Gf=28.28E9; %rhocnt=1300;% E-glass fiber
Vf=Vcnt;


Gm=EC/(2*(1+numm));
Eta1=((Ef/EC)-1)/((Ef/EC)+xi(1));
Eta2=((Gf/Gm)-1)/((Gf/Gm)+xi(2));
%% Effective material properties

E1=Ef*Vf+EC*(1-Vf);
E2=EC*((1+xi(1)*Eta1*Vf)/(1-Eta1*Vf));
G12=Gm*((1+xi(2)*Eta2*Vf)/(1-Eta2*Vf));
G13=G12;
G23=G12;
nu12=nuf*Vf+numm*(1-Vf);
nu21=nu12*(E2/E1);

alpha11f=3.4854e-6;
alpha12cn=5.1682e-6;
alpham=45*(1+0.0005*deltaT)*1.0e-6;
 alp1=(Vcnt*Ef*alpha11f+(1-Vcnt)*Em*alpham)/
(Vcnt*Ef+(1-Vcnt)*Em);
 alp2=(1+numm)*Vcnt*alpha12cn+(1+numm)*(1-Vcnt)*alpham-(nu12*alp1);
 alp12=0;
 [moduli]=[E1 E2 nu12 nu21 G12 alp1 alp2];
V12=nu12;
V21=nu21;
```

Index

For Product Safety Concerns and Information please contact our EU
representative GPSR@taylorandfrancis.com
Taylor & Francis Verlag GmbH, Kaufingerstraße 24, 80331 München, Germany